IMAGES
of Aviation

ALABAMA
AVIATION

With seven aviation training bases located in the state during the Second World War, Alabama served a critical role in the evolution of military airpower as the foundation of Pres. Franklin Roosevelt's arsenal of democracy. At Maxwell Field alone, more than 100,000 cadets were trained as pilots, navigators, and bombardiers. "The road to Tokyo leads through Maxwell Field" became a common theme among cadets. (Air Force Historical Research Agency.)

ON THE COVER: Military aviation cadets prepare for a training flight at the Army Air Forces' advanced flying school at Craig Field in Selma. It was the first specialized single-engine school established in the Southeast, and training at Craig Field was conducted under the jurisdiction of the Southeast Air Corps Training Center at Maxwell Field in Montgomery. (Author's collection.)

IMAGES
of Aviation

ALABAMA
AVIATION

Billy J. Singleton

ARCADIA
PUBLISHING

Published by Arcadia Publishing
Charleston, South Carolina

Library of Congress Control Number: 2017941330

For all general information, please contact Arcadia Publishing:
Telephone 843-853-2070
Fax 843-853-0044
E-mail sales@arcadiapublishing.com
For customer service and orders:
Toll-Free 1-888-313-2665

Visit us on the Internet at www.arcadiapublishing.com

For Patrick Earl Carter, aviator and friend

CONTENTS

ACKNOWLEDGMENTS

The author is deeply indebted to the following individuals and organizations for their contributions and support of this project: Meredith McDonough and the Alabama Department of Archives and History (ADAH); director and staff of the Air Force Historical Research Agency (AFHRA); Air University Office of History (Air University); Maxwell Air Force Base; Alabama Aviation Hall of Fame; Alabama Air National Guard (ANG); Southern Museum of Flight, Birmingham (SMF); National Museum of Naval Aviation; Auburn University Archives; Doy Leale McCall Rare Book and Manuscript Library, University of South Alabama Archives; Historic Mobile Preservation Society; Huntsville-Madison County Public Library; Wright State University; Montgomery Airport Authority; Town of Courtland, Alabama; Thomas Badham; and Anthony Smith, Muscle Shoals Airport.

The author gratefully acknowledges the guidance and assistance of Caitrin Cunningham, senior title manager, Arcadia Publishing.

The author is especially grateful to Lacey Newman for her support, encouragement, and efforts in making this project a reality.

Images not credited are from the author's collection.

INTRODUCTION

Thirteen miles east of the Clarke County community of Grove Hill, a faded blue sign designates the intersection of US Highway 84 and County Road 35. Known as the Old Line Road, County Road 35 traces a path through the thinly populated eastern sections of Clarke County. Located a short distance beyond the end of the paved section of the road, a stone monument commemorates the once thriving settlement of Suggsville. In addition to memorializing this early religious and educational center, the text inscribed on the marker reads, "Site of extensive aviation experiments by Dr. Denny 100 years before Wright Brothers."

The subject of this claim, Andrew Denny (1812–1869), was a physician and naturalist who developed an interest in the concept of a flying machine. Local historians suggest Denny endeavored to construct a machine to be launched from a hill and remain aloft by mechanically replicating the ability of birds to soar through the air. The doctor reportedly retained the services of a skilled mechanic to assist in construction of a machine and invested more than $5,000 in his pursuit of human flight.

Similar to other claims of aerial experiments that predate the Wright brothers, the text relating to the aviation experiments of Andrew Denny is not consistent with other documented events. Born in 1812, Denny could not have conducted extensive aviation experiments 100 years before Orville and Wilbur Wright successfully achieved sustained and controlled flight of a heavier-than-air machine in December 1903 at Kitty Hawk, North Carolina. Even without evidence to substantiate his work, Denny is considered the first experimenter in Alabama history to be associated with the pursuit of human flight.

Although he may have been first, Andrew Denny was not the only experimenter in Alabama to aspire to achieving the goal of human flight. Louis Archer Boswell (1834–1909) of Talladega developed a design that embodied his ideas relating to flight in a heavier-than-air machine. Boswell claimed to have devised all needed devices to steer an aerial machine both left and right and to ascend and descend at will with ease and safety. Unlike his predecessor Denny, Boswell made application and received patents for his "Steering Mechanism for Dirigible Airships and Improvements in Aerial Propeller Wheels."

In Mobile, John Ellis Fowler (1862–1939) was known not only for his extraordinary mechanical ability but also as an experimenter and inventor who developed an interest in the design and construction of a flying machine. A watchmaker by trade, Fowler constructed three prototypes of flying machines. Although he never achieved success with his designs, Fowler did publically display two of his prototypes at Monroe Park in Mobile.

In northern Alabama, acquaintances considered William Lafayette Quick (1859–1927) of New Market an innovator. Like his predecessors, Quick was an astute observer of nature who believed he could devise and construct a flying machine based on his study of ornithology. Quick designed a flying machine that featured an identifiable fuselage that allowed the operator to sit in an upright position. The design also included an innovative landing gear arrangement of wheels instead of skids. Unlike the more common pusher system of propulsion used on other early designs, the Quick machine utilized a tractor propulsion system to pull the machine through the air. In 1908, Quick's son, William Massey Quick, reportedly made one attempt to get the machine airborne although no evidence exists to document the event.

In March 1910, the dream of a flying machine in Alabama became a reality as Orville Wright ascended above the cotton fields west of Montgomery in a Wright Transitional Flyer. The historic flight, the first of a heavier-than-air machine over the state of Alabama, was made in conjunction with establishment of the first civilian flying school in the United States. The school was created by the Wright brothers to train exhibition pilots to demonstrate Wright-produced flying machines to society sportsmen. Orville and Wilbur Wright selected Montgomery as the location of their school because they considered the genial climate and suitable grounds of central Alabama ideal for the operation of their flying machine. At Montgomery, the brothers found not only a favorable environment but also a city and state committed to the development of aviation.

The same factors that attracted Orville and Wilbur Wright to Alabama would become a recurring theme during the early history of powered flight in Alabama. A mild climate that allowed flight training to be conducted throughout the year and the flat terrain of central and south Alabama would be significant factors in attracting military flight training to the state.

The declaration of war against the "Imperial German Government" in April 1917 created an immediate need to train aviators to operate the newest technology in the inventory of the United State military, the airplane. With the rapid expansion of military aviation during the First World War, flight training returned to central Alabama on a much larger scale.

Taylor Field, located southeast of Montgomery in the Pike Road community, was designated one of 14 primary aviation training fields in the Aviation Section of the United States Signal Corps, predecessor to the Army Air Service. Established in November 1917, the 800-acre facility included 16 hangars, repair shops, a hospital, and housing and recreational facilities for personnel assigned to the post. Before the end of the war, approximately 200 training aircraft were assigned to the field.

One of the earliest military aviation training fields in the United States, Taylor Field was also the first airfield of standardized design to be constructed in the state of Alabama. Prior to the First World War, aviators routinely operated flying machines from racetracks, fairgrounds, or any unobstructed surface of sufficient length.

Developed by Albert Kahn, the standard military airfield design used for the construction of Taylor Field consisted of a one-mile-square section of level ground that allowed takeoffs and landings in any direction. The standard design included aircraft hangars along with administration and other buildings situated in a linear arrangement along the flight line.

To support flight training operations at Taylor Field and five other training sites in the Southeast, an engine and repair depot was established near Montgomery. One of only three aviation repair depots in the United States, the Montgomery facility was constructed on the same parcel of land selected by Orville and Wilbur Wright less than a decade earlier as the site of the nation's first civilian flying school. This post would evolve into Maxwell Field and subsequently Maxwell Air Force Base, a facility that has served the citizens of the United States for more than 100 years.

Following the end of the First World War, the military continued to serve a vital role in the development of aviation in Alabama. In 1921, a group of former military pilots organized a flying club in Birmingham to promote the benefits of aviation. James Armand Meissner (1896–1936), a member of the 94th Aero Squadron during the Great War, led the organizing effort.

Members of the flying club made application to obtain designation as a unit of the Alabama National Guard. In January 1922, the group was officially recognized as the 135th Observation Squadron. Work began immediately to develop a flying field located west of the city on a parcel of land donated by the Republic Iron and Steel Company. The field was named for Arthur Meredyth Roberts, an aviator of the Army Air Service who was fatally injured in a training accident in France during the First World War.

Constructed primarily for the use of the National Guard squadron, Roberts Field was also utilized to meet the commercial aviation needs of the city of Birmingham. The field was made available for airmail service and as a stopping point on the proposed dirigible air line from New York to Havana.

Like Roberts Field, Maxwell Field in Montgomery was instrumental in the development of commercial airmail service in Alabama. On April 17, 1925, Capt. Robert D. Knapp and mechanic J.A. Liner flew the first airmail flight into the city of Montgomery. The flight was operated to determine the feasibility of connecting airmail from the Gulf Coast states to the transcontinental service at Chicago.

Flying a de Havilland DH-4 aircraft, Knapp and Liner departed New Orleans for Legion Field in Mobile. After picking up five bags of mail, the pair continued to Maxwell Field. National Guard aviators from Roberts Field completed the next leg of the flight from Montgomery to Nashville via Birmingham.

Airmail service in Alabama was inaugurated May 1, 1928, on Commercial Airmail Route 23 connecting the cities of New Orleans and Atlanta. Operated by St. Tammany Gulf Coast Airways, the route included intermediate stops at Mobile and Birmingham. The service was initially restricted to daylight hours until lighted navigation beacons could be installed at intervals of 15 miles along the route. Intermediate landing fields were constructed at intervals of 30 miles to provide a suitable landing area should an emergency or adverse weather conditions require an unscheduled stop.

In May 1931, as the nation suffered the effects of the economic depression, the city of Birmingham dedicated its new municipal airport. The largest airport in Alabama, the new Birmingham Municipal Airport was the first aviation project in the state to exceed $1 million in construction costs. The facility was also the first to include a paved runway to accommodate scheduled airline service. Recognizing the economic benefits of aviation facilities, other municipalities in Alabama initiated airport construction projects. Civic leaders in Dothan, Huntsville, Mobile, Montgomery, and Tuscaloosa began to develop airport facilities to attract airmail and passenger service to their respective cities.

Ironically, the foundation of the system of airports in Alabama was created during the years of the Great Depression. The Airport Construction Program of the Civil Works Administration (1933–1934) and the Works Progress/Work Projects Administration (1935–1942) was part of an effort to reduce high rates of unemployment created by the financial crisis. These programs resulted in 42 airport projects in 23 communities across Alabama.

An innovative aspect of airport construction during this period was the Alabama Air Park Plan. Created by Sumpter Smith, director of the Civil Works Administration in Alabama, the plan was an attempt to create facilities that served the interests of the entire community. Airports were designed to include recreational amenities such as golf courses and swimming pools for use by the public. Dedicated in 1929, the original Montgomery Municipal Airport was the first to incorporate this concept.

Beginning in 1939, aviation in Alabama would be permanently transformed by the escalation of hostilities that preceded the Second World War. In a worldwide conflict that Pres. Franklin Roosevelt believed would be won by means of superior airpower, the Army Air Corps initiated an effort to create an "arsenal of democracy" built on the foundation of a vast aerial armada.

The same factors that made Alabama an ideal location for the establishment of the nation's first civilian flying school, a genial climate and suitable terrain, would create an environment to make Alabama a leader in producing military pilots for service in the European and Pacific theaters of combat.

In July 1940, the Southeast Air Corps Training Center was established at Maxwell Field. Responsible for all pilot, navigator, and bombardier training at military aviation installations in the Southeast, the center was one of three military aviation training centers in the United States. In May 1942, the center was renamed the Army Air Forces Southeast Training Center before being designated Army Air Forces Eastern Flying Training Command the following year.

From 1939 to 1941, five airfields would be constructed in Alabama to provide basic and advanced pilot training for cadets of the Army Air Corps and its successor, the Army Air Forces: Gunter Field at Montgomery, Napier Field at Dothan, Craig Field at Selma, and the Courtland and Tuskegee Army Air Fields. Additionally, Auxiliary Air Station Barin Field in Baldwin County was utilized to provide training for naval aviators. Primary training for military cadets was provided through civilian contract schools at Tuscaloosa, Decatur, and Tuskegee. Specialized flight training for pilots of military assault gliders was conducted in Mobile.

By 1943, the skies over central Alabama were described as having the densest air traffic in the world due to the massive expansion of military flight training. To reduce aerial congestion at the main bases, more than 30 auxiliary landing fields were constructed across the state. Following the war, many of these military aviation fields would become municipal airports.

In January 1940, a maintenance and repair depot was constructed on the site of the Mobile Municipal Airport to provide logistical support for 31 military airfields in the Southeast and Caribbean areas. Named for Army aviator Wendell Holsworth Brookley, the facility was one of only two military aviation repair depots being utilized in the years immediately preceding the Second World War. Brookley Field was unique in being the only military aviation installation in the United States served directly by four modes of transportation—air, sea, rail, and highway.

Following the end of World War II, military airfields in Alabama were decommissioned as rapidly as they had appeared only a few years earlier. By 1946, flight training at Tuskegee, Courtland, Napier, and Gunter Fields had been terminated and the bases declared surplus property. Brookley Field in Mobile continued to support military aviation operations until its closure in 1966. Craig Field at Selma was decommissioned in September 1977 as part of a reduction of US Air Force training bases nationwide. During 37 years of operation, more than 30,000 pilots were trained at Craig Field.

The end of the Second World War brought to a close a remarkable period in the history of aviation in Alabama. The 100-acre cotton field that was the site of the first flight of a heavier-than-air machine over the state and the nation's first civilian flying school had evolved into a training center having jurisdiction over military flight training installations in the Southeast. Today, Maxwell Air Force Base is the home of the United States Air Force University and the Center for Professional Military Education and Doctrine, the Air Force Research Institute, and numerous professional military schools.

The story of aviation in Alabama began in rural Clarke County as physician and naturalist Andrew Denny envisioned the possibility of applying the principals of nature to the construction of a flying machine. The dream of a flying machine was shared by other early experimenters in Alabama who devoted their lives to the pursuit of mechanical flight. For more than 150 years, aviation and aerospace have remained a source of inspiration and motivation for the people of Alabama.

From early aerial experimenters to partners in the creation of a vast aerial arsenal of democracy during the Second World War, the people of Alabama have demonstrated a firm and enduring commitment to the growth and development of the aviation and aerospace industry. This level of commitment will ensure the legacy of aviation in Alabama as the history of our future is written.

One

THE PIONEERS

Located on County Road 35 east of Grove Hill, a stone marker erected by the Clarke County Historical Society commemorates the settlement of Suggsville. Text inscribed on the marker suggests the early religious and educational center was the site of extensive aviation experiments 100 years before Orville and Wilbur Wright demonstrated the concept of controlled and sustained flight in a heavier-than-air machine.

Dr. Boswell Invents
(of EASTABOGA, TALLADEGA COUNTY, ALABAMA)
An Aeroplane

Man W[...]
Lewis A[...]

Compiled by Vern. Scott for the
Talladega County Historical Association, 1986
This book published by: Joe Upchurch • 735 Glenwood Road • Talladega, Alabama 35160

Born in Lunenburg County, Virginia, Lewis Archer Boswell was known by acquaintances as a medical practitioner whose educational attainments in other fields were of an extraordinary character. In 1869, Boswell moved to Talladega County, where he continued his inventive reflections on the concept of a heavier-than-air flying machine. Before his death in 1909, Boswell secured two patents for his designs. (Talladega County Historical Society.)

In May 1903, Boswell received a patent for his Steering Mechanism for Dirigible Airships. Although the patent specified a control system for dirigible airships, it might have been his intention to combine his concepts for aerial propeller wheels and steering mechanism to create a heavier-than-air flying machine. Boswell devoted the final years of his life soliciting funds for construction of a lightweight motor to propel his design. (U.S. Patent and Trademark Office.)

William Lafayette Quick was born in Hardin County, Tennessee. Although he lacked formal education, Quick was described by acquaintances as an innovator and visionary. After relocating to the community of New Market, located 15 miles northwest of Huntsville, Quick developed an interest in the concept of mechanical flight. An astute observer of nature, Quick believed he could design and build a flying machine that replicated the flight of buzzards. (AAHOF.)

In 1900, William Quick began construction of his flying-machine design, a project that required eight years to complete. The machine utilized a tractor propulsion system unlike the more common pusher designs of the period. Instead of a landing skid arrangement, Quick developed a three-wheeled landing gear. Restored in 1964, the flying machine of William Quick is displayed at the U.S. Space and Rocket Center in Huntsville.

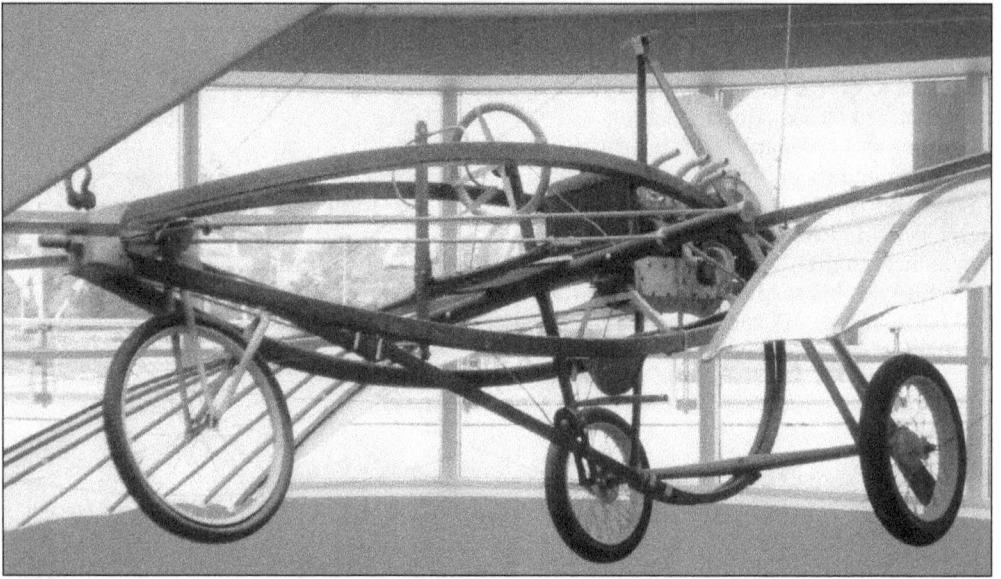

Unlike the biplane arrangement common to designs of the period, Quick developed a monoplane with an identifiable fuselage that allowed the operator to sit in an upright position. The bird-shaped wings were constructed using exposed wooden ribs. The wing fabric, attached to the lower surface, was obtained from a textile mill near Huntsville. For the landing gear, Quick used wire-spoked bicycle wheels.

The original design included a harness mechanism that allowed the operator to control up-and-down movement of the machine. With the harness attached, the operator would lean forward to incline the nose of the machine upward. Positioning the body aft would cause the machine to descend. Quick received a patent in October 1913 for a flying machine that included significant changes to the original design.

Born in Saltillo, Mississippi, John Ellis Fowler relocated to Mobile in 1884 and established a clock and sewing machine repair business on Dauphin Street. Known for his extraordinary mechanical ability, Fowler created three flying-machine designs. Prototypes of two of these designs were constructed behind a large wooden fence on the grounds of Monroe Park in Mobile. (USA Archives.)

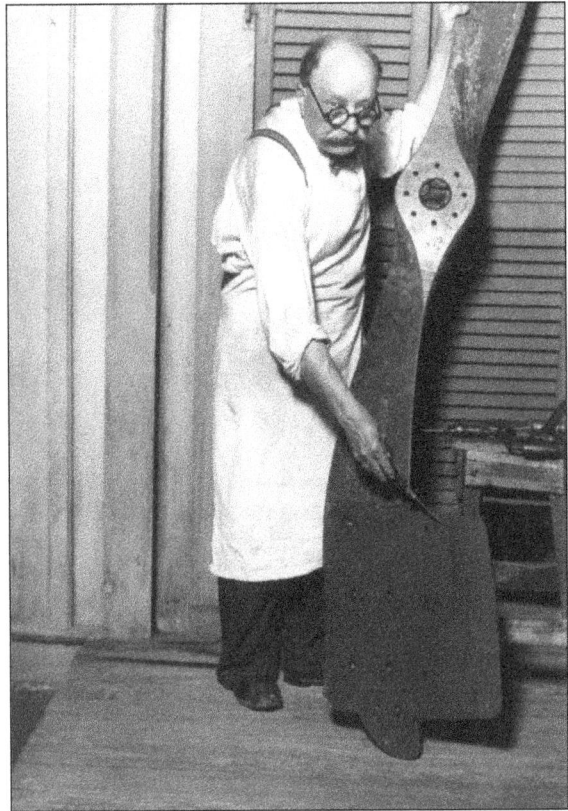

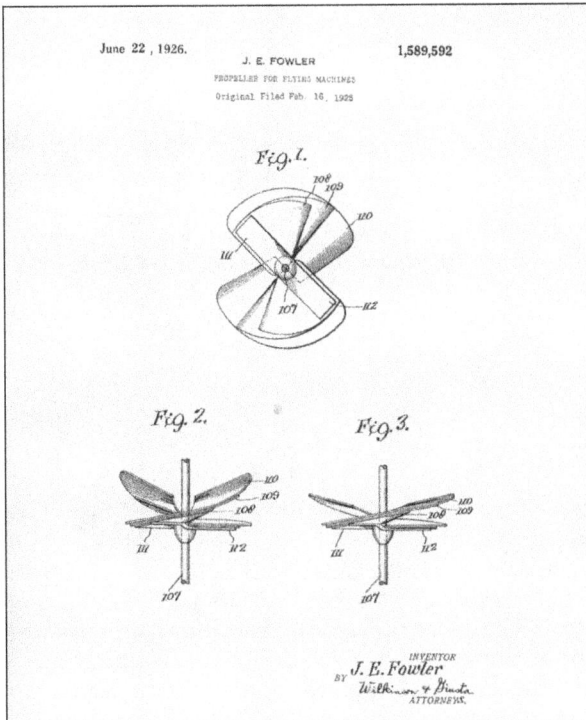

June 22, 1926.   1,589,592
                J. E. FOWLER
        PROPELLER FOR FLYING MACHINES
          Original Filed Feb. 16, 1925

Fig. 1.

Fig. 2.   Fig. 3.

                    INVENTOR
              J. E. Fowler
        BY   Wilkinson & Minala
                    ATTORNEYS.

In June 1922, John Ellis Fowler was awarded two patents. For the first patent, Fowler used his expertise in clock repair to design a mechanism to drive an improved propeller that produced a great pulling effect with less motor speed. The second patent covered his flying-machine concept that was designed to allow ascent and lighting from either the ground or water. (U.S. Patent and Trademark Office.)

15

The first machine constructed by John Fowler at Monroe Park sat on an undercarriage of four bicycle-type wheels. The 1902 monoplane design consisted of one large lifting surface supplemented by a second horizontal plane at the rear of the machine. To obtain financial support for his projects, Fowler charged an admission fee of 10¢ to view the machine and learn about his theories relating to heavier-than-air flight. (Historic Mobile Preservation Society.)

Fowler's second flying-machine design to be constructed at Monroe Park was arranged for ascent and alighting from either the ground or water. The design consisted of three lifting surfaces—the bottom, intermediate, and top planes. The marine-type hull included a retractable landing gear. Fowler was assisted on the project by his brother Robert, shown sitting on the fuselage section in the foreground. (Overbey Collection, USA Archives.)

The headline banner of the *Montgomery Advertiser* newspaper proclaimed "Wizards of Air Plan for Flights in Montgomery." Seeking a genial climate and suitable grounds to train aviators for the Wright Exhibition Company, Wilbur Wright selected a parcel of land located only a few miles from the state capital at Montgomery. Recognizing the commercial benefits of being associated with the school, merchants adorned the exterior of the hangar with advertisements. (Wright State University.)

On March 26, 1910, Orville Wright ascended above the cotton fields west of Montgomery in a Wright Transitional Flyer; it was the first flight of a heavier-than-air flying machine over Alabama. During the flight, Wright remained aloft for five minutes and achieved an altitude of approximately 50 feet. Because flying activities were not scheduled to begin until the following week, few spectators were present to witness the historic event. (Wright State University.)

Walter Brookins (left) was not only the first student aviator in Alabama history but also the first to be trained as a flight instructor. With less than five hours of experience in the air, Brookins learned to operate the Flyer from the right-hand seat as an instructor. In May 1910, Brookins and student Archibald Hoxsey (right) completed one of the first night flights in aviation history at Montgomery. (Wright State University.)

During the early 20th century, the airplane became a symbol of innovative thinking. Local civic leaders took advantage of the presence of the Wright brothers to promote Montgomery as a progressive city. This vintage postcard illustrates an exceptionally accurate depiction of the Wright Flyer over Court Square in Montgomery—even though the Wright Flyer was never operated over the downtown area of the city.

Beginning in 1911, flying machines began to appear at aerial exhibitions, state fairs, and other venues in Alabama. The state fairgrounds in Birmingham, Vandiver Park in Montgomery, and Monroe Park in Mobile were sites often used by exhibition teams and individual aviators to thrill crowds with death-defying aerial stunts. (SMF.)

In October 1911, California native Robert Fowler departed Los Angeles in a successful attempt to complete the first transcontinental flight from west to east. Piloting a Wright Model B Flyer, Fowler landed in Mobile in January 1912 before departing for Flomaton. In Alabama, Fowler also landed at Flomaton, Georgiana, Brantley, and Troy. Because of an accident on takeoff, Fowler spent two weeks at Brantley repairing his machine. (Wright State University.)

Born at Webster, Michigan, Osbert Edwin Williams (1875–1917) became one of the most successful flyers in America. In 1916, Williams relocated his family to a farm near St. Elmo in Mobile County to take advantage of the mild weather conditions that would allow flying activities throughout the year. The following year, Williams lost his life during an aerial exhibition at the Gulf Coast Fairgrounds. (Nancy Mess.)

Born in Fort Payne, Katherine Stinson (1891–1977) became one of the pioneers in aviation history. Known as the "Flying Schoolgirl," Stinson became the fourth female pilot in the United States to earn a license to operate an aircraft. In 1914, she became the first female aviator to deliver mail by air. In 2001, Stinson was inducted into the Alabama Aviation Hall of Fame.

Two

THE GREAT WAR

Aviator Ruth Law was the first female authorized to wear a military uniform as a noncommissioned officer. After being denied permission to participate in combat, Law supported the war effort touring the United States and dropping Liberty bond pamphlets from her flying machine. She also raised money for the Red Cross by performing exhibition flights. In October 1917, Law performed aerial stunts for troops stationed at Camp McClellan in Calhoun County.

Activated in December 1917, Taylor Field was located seven miles southeast of Montgomery. The 800-acre facility was one of 14 primary training fields operated by the Aviation Section of the United States Signal Corps. The field was named for Capt. Ralph Taylor of Stamford, Connecticut. Taylor was a military aviator who lost his life in August 1917 in an aviation accident at Mineola Field in New York. (Air University.)

Taylor Field was the first specifically designed flying field to be constructed in the state of Alabama. In May 1917, Albert Kahn was commissioned by the Signal Corps to produce a standard airfield design, the Mobilization Hangar Plan. The design consisted of a one-square-mile section of land that included a flying field, aircraft maintenance and storage hangars, and administration and other buildings to support 100 aircraft and 150 student aviators.

Taylor Field was reportedly one of the most ideal training facilities in the United States. Army inspectors described the flying field as being "level as a floor." The mild climate allowed training to be conducted throughout the year. The gravel runways visible in the image were the first to be constructed in Alabama and were utilized during periods of wet weather that made portions of the flying field unusable. (Air University.)

From May 1918 until April 1919, more than 200 Curtiss JN-4 and de Havilland DH-4 training aircraft were operated at Taylor Field. The flight training program included instruction in basic flight maneuvers, cross-country operations, formation flying, and aerobatics. Successful completion of the Reserve Military Aviator examination at the completion of training entitled the cadet to wear the wings of a military aviator. (Air University.)

The open cockpit arrangement of training aircraft at Taylor Field exposed students and instructors to extremes of temperature and wind. For protection, standard aviator attire included a heavy leather coat, gloves, boots, helmet, and goggles. This equipment protected the aviator not only from the elements of weather but also from the oil, grease, and gasoline discharged from the engine during flight.

The flight training program included approximately 50 hours of flight instruction. After demonstrating proficiency in basic maneuvers, cadets progressed to the cross-country phase of training that included flights around a triangular course of a specified distance. Rudimentary navigational charts and a lack of prepared emergency landing fields made this one of the most demanding phases of the training program.

TAYLOR FIELD

Montgomery's first military flying installation was established 200 yards south of this spot in November of 1917. The facility was named for Captain Ralph L. Taylor, who was killed in an airplane crash at Mineola Field, New York in August of 1917. The primary flying school here included 16 hangars, repair shops, warehouses, quarters, a hospital, and nearly 200 JN-4 and DH-4 aircraft on its 800 acres. One hundred and thirty-nine fledgling pilots completed the eight-week course and some served in France during the First World War. Taylor Field closed in April of 1919 and reopened as Gunter Auxiliary Air Field No. 5 during World War II. It was closed again in July of 1946.

Before it was deactivated in April 1919, one hundred thirty-nine cadets graduated from the training program at Taylor Field. The landing area was again utilized during the Second World War as Gunter Field Auxiliary Landing Field No. 5. The field was permanently deactivated in July 1946 and the land returned to the original owners. The only remnants of the first military flying facility in the state of Alabama are the concrete hangar pads and swimming pool. In 1993, a marker was erected by the Montgomery Chapter of the Air Force Association and Founders of Flight, Order of Daedalians to commemorate the facility.

General View "Wright Field". Aviation Repair Depot #3. Montgomery, Ala. Jan. 9th 1919.

PHOTO #361
Paul L. Richards.
Montgomery. Ala.

In April 1918, inspectors of the Aviation Section of the US Signal Corps, predecessor to the Army Air Service, selected Montgomery as the site of a repair depot designated to provide support for training operations at Taylor Field and five other training schools in the Southeast. The 302-acre tract of land selected included the site of the nation's first civilian flying school, operated by Orville and Wilbur Wright.

The maintenance workshops of the depot were equipped to produce or repair virtually every component of the training aircraft of the United States Army. The Montgomery Engine and Repair Depot supported flight training operations at Taylor Field; Gerstner Field, Lake Charles, Louisiana; Payne Field, West Point, Mississippi; Souther Field, Americus, Georgia; and Door and Carlstrom Fields in Arcadia, Florida.

Less than 90 days after construction began, 52 buildings that included barracks, hangars, a commissary, post exchange, and infirmary were completed. The post had many modern conveniences that were not routinely available to residents of Montgomery County, including electricity and telephone service. In September 1918, a flag-raising ceremony marked the official opening of the post, one of three aviation repair depots in the United States.

The landing area of the Montgomery depot consisted of a sod surface. Because there were no designated runways, pilots could align their aircraft into the prevailing wind for takeoffs and landings. Although it was a military facility, transient pilots were allowed to use the field. In May 1923, aviator Charles Lindbergh landed at the depot to repair the radiator of his recently purchased military surplus Curtiss JN-4 "Jenny" aircraft. (ADAH.)

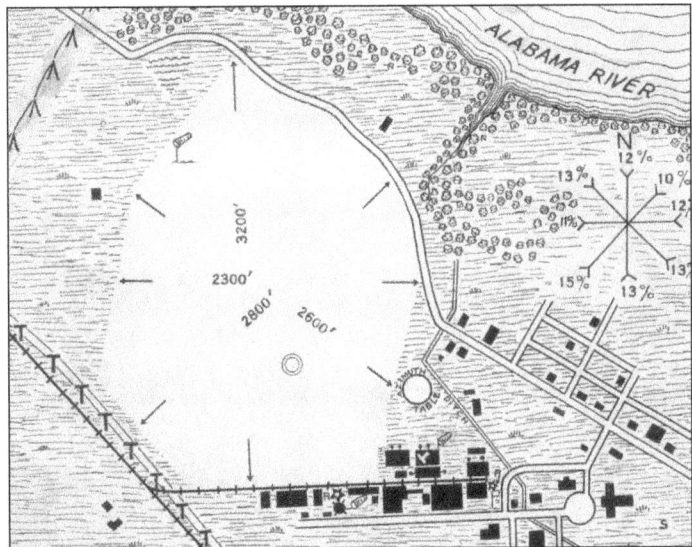

The Honorable Charles Henderson, governor of the state of Alabama, attended the dedication ceremony of the Aviation and Repair Depot. After addressing the assembled crowd, Henderson (right) was taken aloft by pilot Ross L. Smith (left) in a Curtiss JN-4 aircraft. During the brief flight over the city of Montgomery, Henderson became the first Alabama governor to go aloft in an airplane. (ADAH.)

In September 1918, the Montgomery depot was renamed the Engine and Plane Repair Depot No. 3. In March 1919, the field was designated the Aviation Repair Depot-Montgomery, or ARDMONT. In 1921, the facility became the Montgomery Air Intermediate Deport. In November 1922, the base was named for military aviator William Calvin Maxwell (1892–1920) of Natchez, Alabama, who was fatally injured in an aircraft accident in the Philippines.

Charles Rudolph d'Olive (1896–1974) was born in Suggsville, Alabama. He enlisted in the military in April 1917 as the United States declared war against the German Empire. Trained as an aviator in France, d'Olive was assigned to the 93rd Aero Squadron. He earned a Distinguished Service Cross by downing three enemy aircraft in one day. During the war, d'Olive was credited with the destruction of five enemy aircraft.

William Terry Badham (1895–1991) was born in Birmingham. Following graduation from Yale, Badham joined the French Air Service. In August 1918, he transferred to the 91st Aero Squadron of the Army Air Service. Shown in the cockpit of a Salmson 2A2 aircraft, Badham was credited with the destruction of five enemy aircraft as an observer during the First World War. Badham retired from military service as a brigadier general. (AAHOF.)

Henry Lee Badham Jr. (1892–1978) volunteered for military service after graduating from Yale. Badham completed flight training at Tours, France, flying solo after less than two hours of instruction. After receiving a commission in the French Air Service, Badham was transferred to the Third Training Center at Issoudun, France. Because of his excellent flying skills and leadership abilities, Badham was promoted to chief instructor of the training facility. Following the war, he returned to Alabama and was instrumental in organizing the Birmingham Flying Club that would evolve into the 106th Observation Squadron. Badham served with distinction during the Second World War and retired as a brigadier general. In 2007, he was inducted into the Alabama Aviation Hall of Fame. (Both, Thomas Badham.)

Three

THE GOLDEN AGE

A native of Anniston, Ruth Elder (1902–1977) was known as the "Miss America of Aviation." Elder became the first female aviator to attempt to cross the Atlantic Ocean by air. She and her copilot were forced to make an emergency landing at sea because of an engine oil leak. At the time, the October 1927 attempt represented the longest nonstop flight by a female aviator.

James Armand Meissner (1896–1936) was born in Londonderry, Nova Scotia, Canada. During the First World War, Meissner served as a member of the 94th Aero Squadron in France and was credited with the destruction of eight enemy aircraft. After the war, Meissner led the effort to organize a National Guard aviation unit in Birmingham and was instrumental in the establishment of Roberts Field. (AAHOF.)

In 1919, the Birmingham Flying Club was formed to promote aviation in the city. Members of the club immediately began to petition the federal government to recognize the unit as a National Guard aviation squadron. In January 1922, the organization was officially designated the 135th Observation Squadron, Alabama National Guard. (ANG.)

On a parcel of land donated by the Republic Iron and Steel Company, the 135th Observation Squadron established a flying field located west of the city. The field was named for Lt. Arthur Meredyth Roberts, a military aviator from Birmingham who died in France during the First World War.

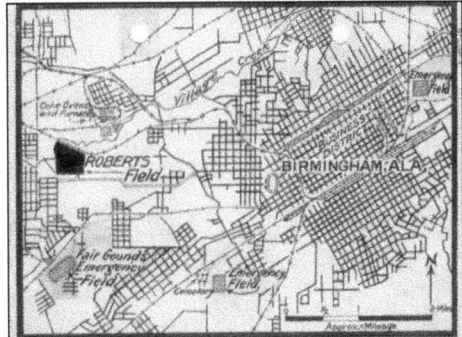

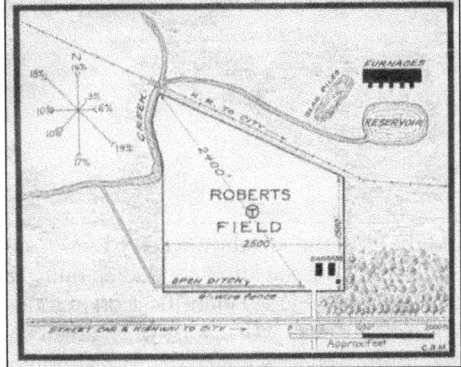

Although developed as a military installation, Roberts Field also served as the first municipal airport in Birmingham. This Department of Commerce *Airway Bulletin* provided important information relating to the airfield for transient pilots. Note the existence of emergency landing fields to the southwest, south, and east. The furnaces and slag piles of the steel mill located adjacent to the field created obstructions in the landing and takeoff path. (ADAH.)

In January 1923, the 135th Observation Squadron was redesignated the 114th Observation Squadron. In January 1924, the designation was changed to the 106th Observation Squadron. The early years of the organization proved to be financially challenging. Members of the squadron often declined salaries to provide additional funding for improvements to the facility.

In July 1922, the squadron received seven surplus Curtiss JN-4D aircraft and support equipment. Members of the squadron conducted numerous aerial mapping and survey flights for government projects that included identification of routes for highways as well as waterways and locations for the construction of dams. In 1925, members of the 106th Observation Squadron served a vital role in the development of the first airmail routes in Alabama. (ANG.)

Aviators of the 106th Observation Squadron stand in front of de Havilland DH-4 observation aircraft during a summer encampment at Legion Field in Mobile. In March 1924, after two years of operation, an inspection conducted by personnel of the US Army rated the squadron among the very best in the nation. (ANG.)

During the formative years, the 106th Observation Squadron operated open cockpit aircraft no longer utilized by the Army. These aircraft included only basic instrumentation to assist the pilot during flight. In addition to instrumentation for oil pressure and temperature, a water temperature gauge was installed to monitor operation of the liquid-cooled engine. A rudimentary electrical system provided power for the aircraft landing light. (Thomas Badham.)

The original officers of the 106th Observation Squadron had previously served as military pilots, while a majority of the enlisted men were mechanics with experience at aviation repair depots. To recruit new members, the squadron published advertisements in local newspapers. The 106th and its successor units continue to serve the people of the United States after more than 90 years of operation. (Thomas Badham.)

This Curtiss JN-4 "Jenny" aircraft from Roberts Field is shown flying over the city of Birmingham. Note the external bracing wires that provided strength to the wing structure. Pilots were able to approximate the flying speed of the aircraft by listening to the sound made by the flying wires. (Thomas Badham.)

Pilots of the 106th Observation Squadron were available for immediate activation to military service in times of national emergency. Consisting primarily of pilots that left the service following the First World War, these aviators would be instantly available with minimal additional training. For individual pilots, the National Guard provided the opportunity to enjoy the thrill of flying at government expense. (ANG.)

IN MEMORY OF
DENNIS O. GABBERT
OCT. 26, 1901
MAY 23, 1926

WHOSE PLANE FELL FROM AMONG THE
CLOUDS CARRYING HIS MANLY FORM
TO THIS SPOT OF EARTH HIS SOUL WENT
BACK TO GOD WHO GAVE IT

WITH THE CRUSHING PLANE HIS MANLY
FORM CAME WHIRLING TO THE GROUND
HE FOUGHT SO HARD TO SAVE HIS LIFE
BUT UTTERED NOT ONE SOUND ALAS
THE ANGELS THEY DID COME AND TAKE
HIM HOME TO GOD ALL WE HAVE LEFT
IS HIS PRECIOUS FORM BENEATH THE SOD

Located on the grounds of Bud Newell Park in Hueytown, a marble monument was erected in memory of Sgt. Dennis O. Gabbert, a member of the 106th Observation Squadron at Roberts Field. On May 23, 1926, Gabbert and passenger George Bice were fatally injured after their Curtiss JN-4 "fell from among the clouds." The marker originally included a medallion photograph of the flier seated in his machine.

Roberts Field would serve an important role in supporting commercial aviation operations for the city of Birmingham. On May 1, 1928, St. Tammany Gulf Coast Airways inaugurated airmail service to the city as part of Commercial Airmail Route 23 from New Orleans to Atlanta. Birmingham native Vivian Jones piloted one of the inaugural flights. (USA Archives.)

Because of the heavy volume of mail scheduled for the inaugural flight, two aircraft were operated into Roberts Field, a Travel Air and Fokker Universal. The cost of airmail delivery was 10¢ per each half ounce, the weight of two sheets of paper in a business envelope. Initially, St. Tammany Gulf Coast Airways operated one northbound and one southbound flight each day on the route.

In October 1927, Roberts Field received one of its most distinguished visitors. Conducting a nationwide tour to promote aviation after his successful New York–Paris transatlantic flight, Charles Lindbergh was greeted by massive crowds hoping to catch a glimpse of the Lone Eagle. The *Spirit of St. Louis*, the Ryan monoplane flown by Lindbergh, was stored in one of the hangars on the field during his stay in Birmingham. (USA Archives.)

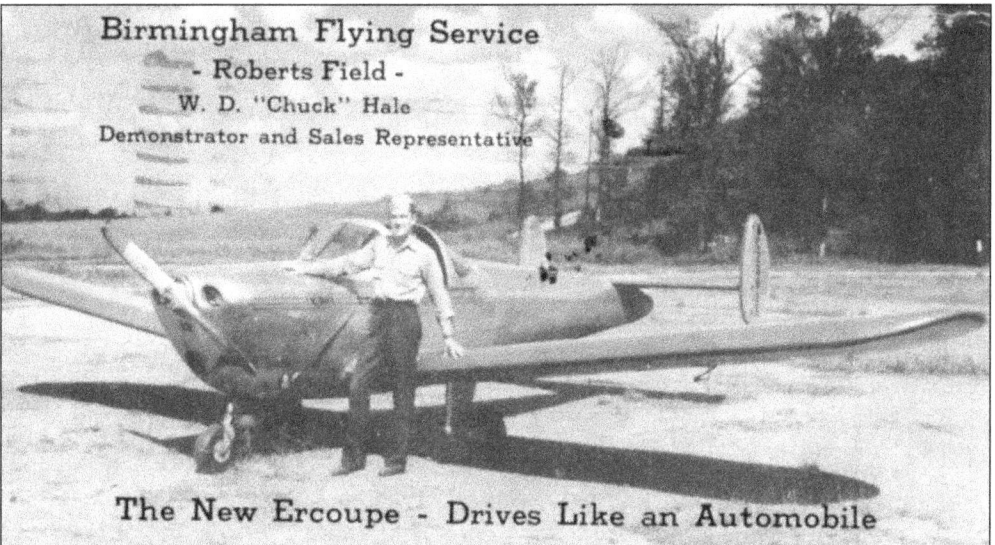

Ironically, the success of commercial aviation in Birmingham was the principal factor in the demise of Roberts Field as a municipal airport. The field was considered too small for the safe operation of larger aircraft being used to accommodate increasing levels of passengers and mail. After the new Birmingham Municipal Airport opened in 1931, Roberts Field served as a general aviation facility until being closed in 1950.

Glenn Messer (1895–1995) was an aviation institution in Alabama. Born in Iowa, Messer learned to fly at 16 years of age. During the First World War, Messer served as a flight instructor at Kelly Field, Texas. He later moved to Birmingham and formed a flying circus that performed over cities in the Southeast. In 1981, Messer was a member of the first class to be inducted into the Alabama Aviation Hall of Fame.

In March 1925, Glenn Messer pioneered an airmail route from Birmingham to Chattanooga, Tennessee. Messer made the flight in a military surplus Curtiss JN-4 that had a maximum cruising speed of 75 miles per hour and limited endurance. With no navigation aids or improved landing areas, the flight was a significant achievement. To commemorate the 50th anniversary, Messer repeated the flight at 79 years of age.

Glenn Messer and fellow Birmingham aviator Virgil Evans cofounded one of the earliest flying fields in Alabama. Located in the western part of Birmingham adjacent to Elmwood Cemetery, the flying field was originally known as Dixie Field. The name would be changed in 1927 to Messer Field in recognition of the founder's increasing notoriety as an aviator. In 1935, the field was renamed Central Park Airport. (ADAH.)

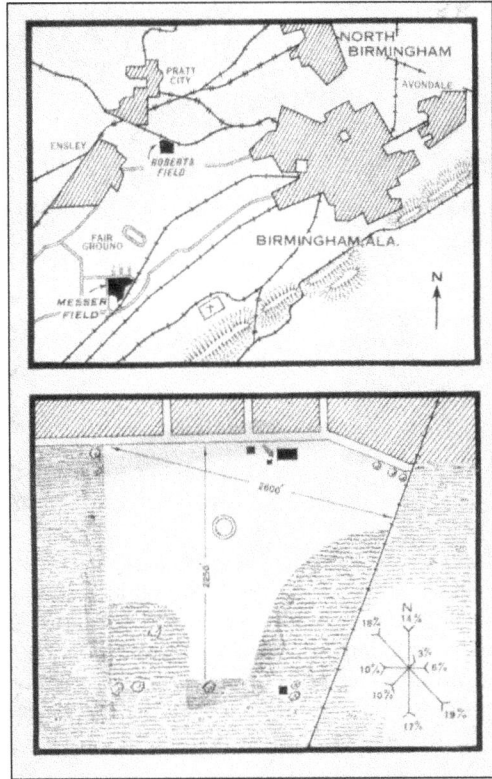

In 1926, the Southern Aircraft Corporation Air Boss was introduced to the flying public. Designed by Glenn Messer and produced until 1930, the Air Boss was advertised as a ship of rugged stamina, great comfort, and maximum safety—a ship of proven mechanical perfection: "When you leave the ground behind you in an Air Boss, you soar away with a certain new-born sense of air freedom."

Following the end of the First World War, the availability of surplus military aircraft made it possible for entrepreneurs to establish aviation services with minimum investment. The Alabama Aviation Company utilized a Curtiss JN-4 to provide flight instruction and sightseeing flights to adventurous residents of Bessemer. Before the development of municipal airports, many of these businesses operated from local farm fields leased from the owner.

The New York–Paris transatlantic flight by Charles Lindbergh in May 1927 generated considerable enthusiasm for aviation in Alabama. Described by the news media as "air minded," young men and women devised ways to enjoy the popular pastime of flying. University of Alabama students utilized a primary glider, towed aloft by an automobile, as an inexpensive means to fly.

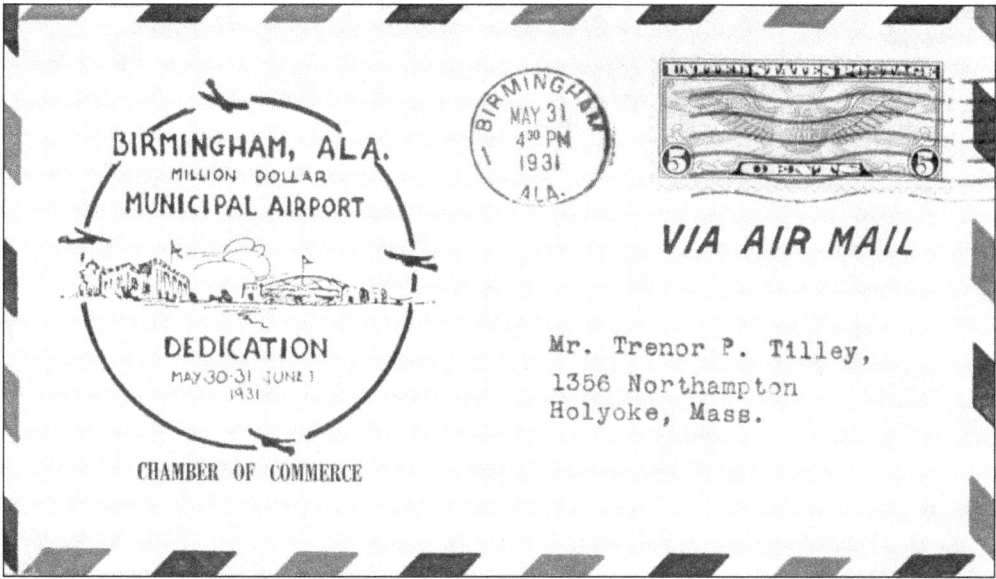

In February 1929, the Junior Chamber of Commerce of Birmingham produced a report promoting the development of a new municipal airport in the city. *An Estimate of the Immediate Need for a Municipal Airport* illustrates the inadequate infrastructure of Roberts Field in meeting the commercial aviation requirements of the region. Members of the city commission authorized a special election to approve bonds to build the new airport.

The new Birmingham Municipal Airport was dedicated during the final weekend of May 1931. The first aviation project in Alabama to exceed $1 million in construction costs, the facility became known as "Alabama's Million Dollar Airport." The field was also the first in the state to utilize a paved runway. The facility was located only minutes from downtown by motorcar on paved thoroughfares.

[V-6541-704B-106)(12-6000)(11-1-39-1P] Municipal Airport Birmingham Ala.

This 1939 aerial view of the Birmingham Municipal Airport shows the 3,700-foot northeast-southwest and the 3,600-foot north-south runways. American Airlines required construction of paved runways to continue service to the city. The operating area of the 106th Observation Squadron on the north side of the airport is identified by the checkerboard pattern of the hangar roof in the upper left section of the image.

This 1939 Department of Commerce *Airport Data Sheet* provided information for operators of aircraft using the Birmingham Municipal Airport. The 315-acre facility included a restaurant and overnight accommodations. For night operations, the airport was equipped with boundary and obstruction lighting. Routine servicing of aircraft was available night and day.

A concrete parking apron and taxi strips located adjacent to the hangar and administration building were provided to eliminate dust from surface wind and aircraft propeller blasts. The name of the airport was painted on the roof of the hangar to assist pilots in determining their geographical position. The letter M displayed above the airport name identified the facility as a municipal airport.

The hangar of the Birmingham Municipal Airport included space for offices and storage on each of the four corners of the structure. The airport wind sock was attached to the northeast corner of the hangar. Overnight storage rates were determined by the number of motors on the aircraft. Licensed aircraft mechanics were on duty night and day.

The architectural theme of the terminal and administration building was Southern Colonial. Beautiful columns adorned the front of the structure. The building included waiting rooms, a ticket office, dining room, dormitories, and other comforts. Upstairs observation decks provided panoramic views of the landing field.

The parking lot of the airport terminal provided ample space for visitors and overnight guests. The spare tire cover of the automobile in the image advertises "Learn to Fly, Birmingham Air Service." The sign to the right of the terminal building advertises the availability of airplane rides. For individuals desiring to spend the afternoon watching airplanes take off and land, an outside patio area with concrete benches was available.

Located on top of the terminal building, the air traffic control tower provided services to pilots of aircraft using the Birmingham Municipal Airport. In an era prior to the introduction of air traffic control radar, control tower personnel provided only local weather conditions and traffic advisories to transient pilots.

In the finest southern tradition, rocking chairs were placed on the front porch of the terminal building to allow departing passengers a relaxing way to pass time waiting for their flights. The inauguration of the southern transcontinental air route operated by American Airlines linked Birmingham with Atlanta to the east and Fort Worth, Texas, to the west.

For many years, Steadham Acker (1896–1952) of Birmingham was the foremost producer of air shows in America. Born in Gadsden, Acker served as an aviation instructor in the US Navy during the First World War. Acker served as the first manager of the Birmingham Airport. In 1984, he was inducted into the Alabama Aviation Hall of Fame.

The first Birmingham Air Carnival was held in conjunction with the dedication of the new municipal airport. In subsequent years, the event would emerge in prominence comparable to the Tournament of Roses Parade, Indianapolis 500, and Miss America beauty pageant. The event demonstrated to the rest of the nation the meaning of real southern hospitality. The air carnival was created by Steadham Acker.

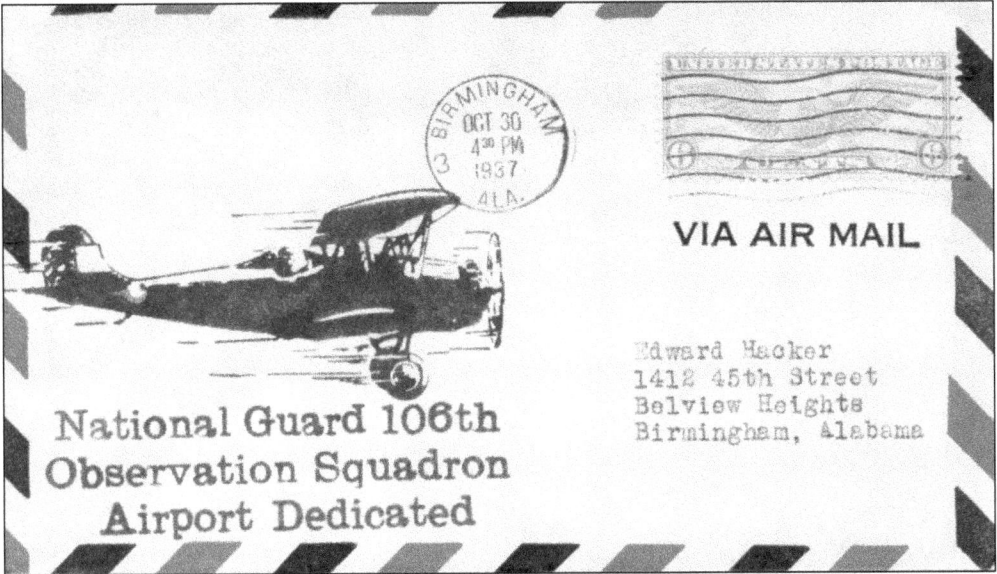

National Guard 106th Observation Squadron Airport Dedicated

VIA AIR MAIL

Edward Hacker
1412 45th Street
Belview Heights
Birmingham, Alabama

Advances in aviation technology and aircraft performance ultimately made Roberts Field unsuitable for the operation of modern military aircraft. In 1935, the Works Progress Administration began construction of a new facility for the 106th Observation Squadron at the Birmingham Municipal Airport. Dedicated In October 1937, the new facility was named for Walter Sumpter Smith, who had been instrumental in the construction and development of the Birmingham Municipal Airport.

Walter Sumpter Smith (1896–1943) served as an instructor in advanced pursuit aircraft during the First World War. In 1922, Smith was instrumental in organizing the 135th Observation Squadron at Roberts Field. In 1935, he was appointed director of the Airways and Airports Division in Washington and was responsible for the design and construction of more than 600 airports nationwide. In 1984, Smith was inducted into the Alabama Aviation Hall of Fame.

49

(C-6044-7048-106X10-12-37.10AX12-200). Squadron Buildings, Municipal Airport.

Construction of the new Birmingham Municipal Airport created an opportunity for the 106th Observation Squadron to expand and modernize its facilities. Construction of aircraft hangars as well as administration and other support buildings was funded through the Works Progress Administration. The unit relocated from Roberts Field to the Birmingham Municipal Airport in 1938 and has remained at that location for eight decades.

The new hangar facilities of the 106th Observation Squadron were designed to accommodate and support North American 0-47 observation aircraft. In November 1940, the squadron was called to active duty in response to increasing hostilities overseas. On December 12, 1941, five days after the Japanese attack on Pearl Harbor, the 106th Observation Squadron was deployed to Miami to perform antisubmarine patrols over the Atlantic Ocean and Gulf of Mexico.

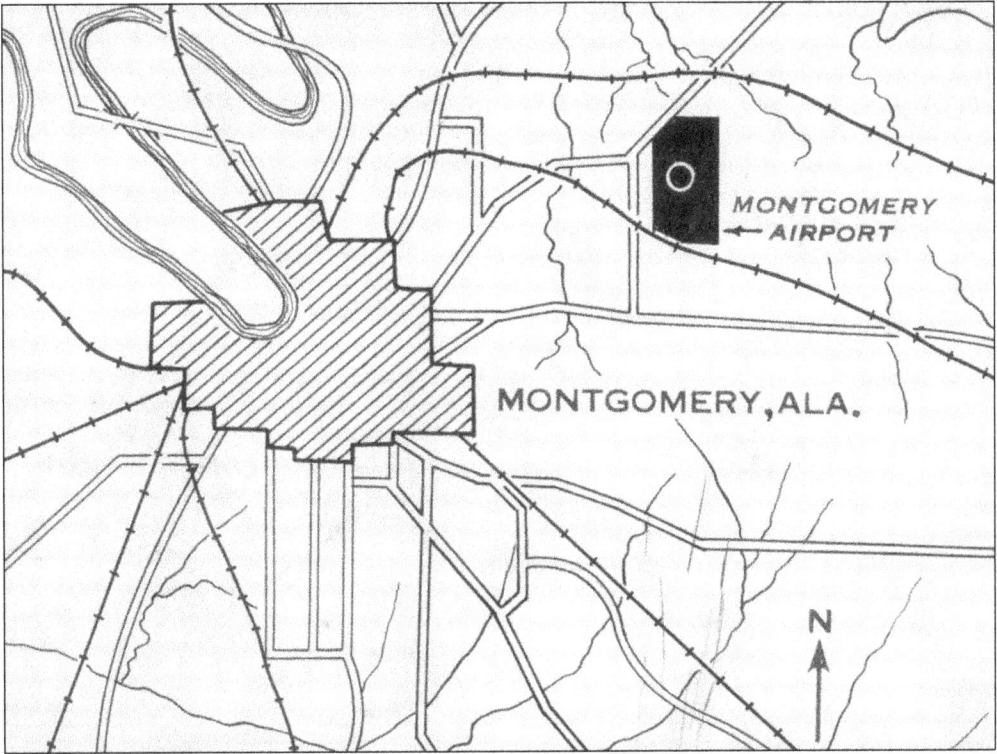

Determined to develop commercial aviation in the city of Montgomery, Mayor William Gunter believed construction of a municipal airport would attract commercial airline service to the city. With inauguration of the southern transcontinental airmail route in 1932, the city of Montgomery replaced Birmingham as an intermediate stop on Commercial Airmail Route 23 from New Orleans to Atlanta. (ADAH.)

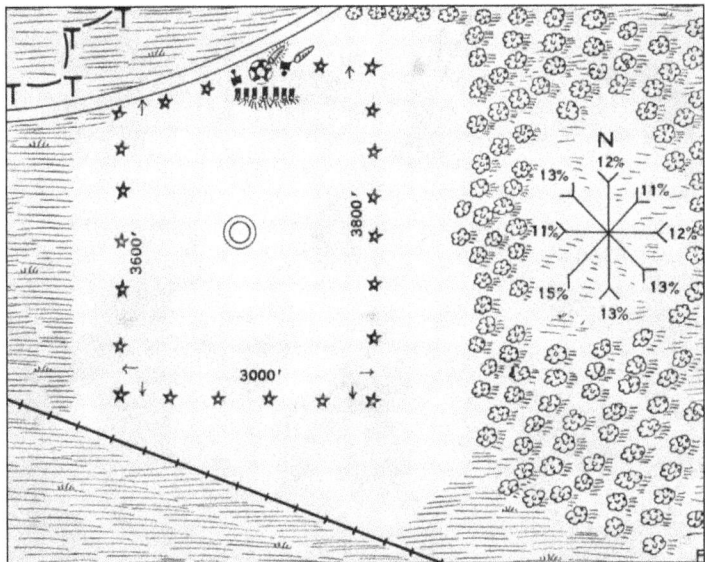

The 600-acre municipal airport consisted of a sod landing area with a white circular field marker located in the center of the landing area. Beacon, boundary, obstruction, and landing-area floodlights for night operations were available on request. The available landing distances are depicted on each side of the landing field graphic. A prevailing surface wind chart is located to the right of the airport diagram. (ADAH.)

The municipal hangar was located on the northern edge of the field. A two-story brick-veneer building situated adjacent to the hangar served as a hotel and restaurant for transient pilots and passengers. The location of the airport relative to the nearest city is prominently displayed adjacent to the administration building to assist transient pilots in verifying their position. (Air University.)

The Manager's Room of the new municipal airport was tastefully furnished to provide pilots and other visitors with comfortable accommodations to await the arrival or departure of their flights. The swimming pool was located to the rear of the building. On the second floor, rooms were available for transient pilots and passengers to secure comfortable accommodations for the night.

In accordance with the Alabama Air Park Plan, the Montgomery Municipal Airport was designed to combine aeronautical and recreational facilities in a central location to serve the entire community. An innovative program created by Sumpter Smith, director of the Civil Works Administration, the Alabama Air Park Plan promoted the development of aviation facilities that included golf courses, swimming pools, and other recreational activities for use by the public. Dedicated in 1929, the original Montgomery Municipal Airport was the first airport in Alabama to incorporate this concept. In the image above, the administration building and hangar are visible to the rear of the caddy house. (Above, ADAH; below, AFHRA.)

B-1440 MUNICIPAL AIRPORT, MONTGOMERY ALA., (9-9-29-1:40P)(12-800)

The Montgomery Municipal Airport was located three miles northeast of the city on Upper Wetumpka Road. Early promotional literature for the city suggested that while aviation had not yet proved practical for the purposes of general transportation, airplane travel was rapidly growing in popularity. Montgomery was no longer considered merely a city of Alabama or the South, but a progressive cog in the giant wheel of America.

B-1441 MUNICIPAL AIRPORT, MONTGOMERY ALA.,(9-5-29-1:45P)(12-800)

At what was pronounced by experts as one of the best all-weather landing fields in the country, American Airways (predecessor to American Airlines) operated two flights each day into Montgomery. Northbound service connected Alabama's capital city with Atlanta, while passengers traveling south were able to fly directly to New Orleans via Mobile. Early advertisements advised prospective passengers that travel by air included all of the comforts and conveniences of home.

54

This October 1929 image of the Montgomery Municipal Airport shows a recently constructed automobile service station on Upper Wetumpka Road adjacent to the field. Civic leaders anticipated construction of the airport would, in addition to attracting local business development, position the city as the aviation center of the South.

This 1941 aerial view of the Montgomery Municipal Airport shows the macadam runway that had been completed three years earlier. The 3,500-foot runway was oriented in a northwest-southeast direction. The need for a hard-surface runway became evident when airline companies refused to operate new Douglas DC-3 aircraft on the sod surface of the airport during periods of inclement weather. (AFHRA.)

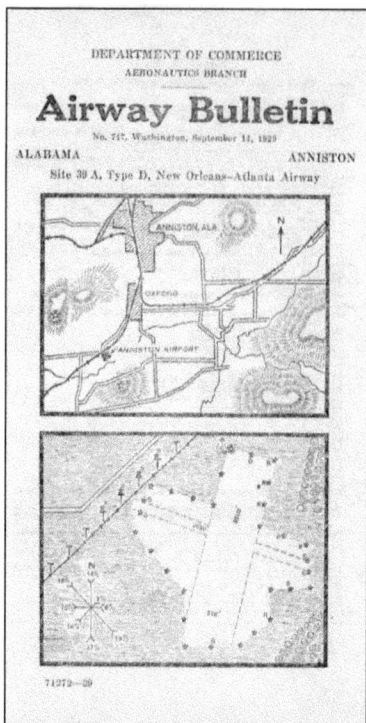

DEPARTMENT OF COMMERCE
AERONAUTICS BRANCH

Airway Bulletin

No. 717, Washington, September 11, 1929

ALABAMA   ANNISTON

Site 39 A, Type D, New Orleans–Atlanta Airway

The September 1929 Department of Commerce *Airway Bulletin* for Anniston, Site 39-A on the New Orleans–Atlanta airway, indicates the airfield was located five and a half miles south of the Anniston Post Office. The field had two sod runways, each 2,000 feet in length. The Type D airport rating related to the lighting aids available for night operations.

The inauguration of commercial airmail service created the need for a system of airports in Alabama. Mail transfer points were established at airports in Mobile and Birmingham. In addition to cities served on the route, a series of landing fields that included the Anniston Intermediate Field were constructed at intervals along the route to provide suitable landing areas should inclement weather or a mechanical problem require an unscheduled landing.

Milton "Skeets" Elliott (right) is shown with S.J. Short (left) and Ormer Locklear. A native of Gadsden, Elliott (1894–1920) trained as a military aviator during the First World War. He completed flight training before he learned to drive a car. Inseparable friends, Locklear and Elliott secretly perfected a series of aerial stunts while serving as pilots in the Army Air Service. (Library of Congress.)

Milton "Skeets" Elliott pilots a Curtiss JN-4 while Ormer Locklear hangs from the landing gear during a performance of one of their death-defying aerial acts. The duo of Locklear and Elliott performed at fairs and other community venues. The men gained notoriety as performers in the silent-movie industry before being fatally injured filming an aerial stunt for the movie *The Skywayman*. (University of Texas.)

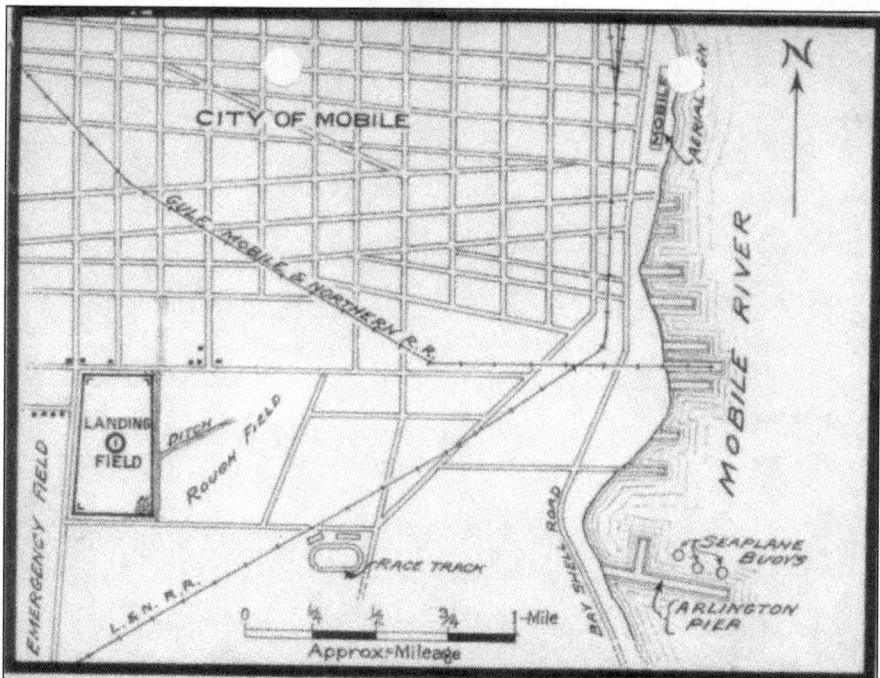

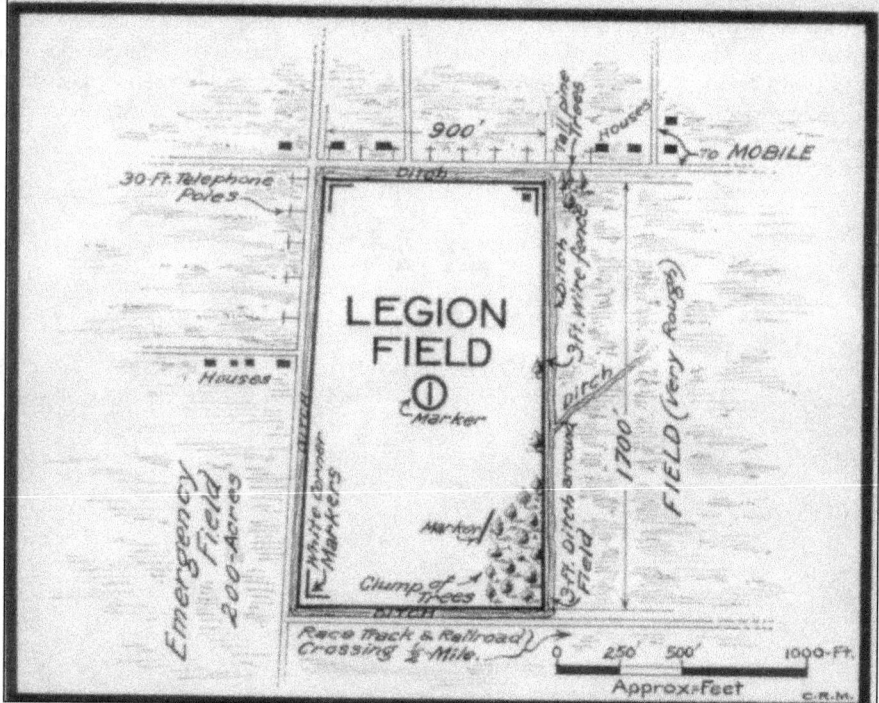

The first municipal airport to serve Mobile was on a 100-acre tract of land located at the south end of Ann Street extending for a number of blocks along Duval Street. The parcel of property was owned by Mobile attorney Harry Toulmin Smith, who leased it to the city in 1917 for use as an airfield. The landing area of Legion Field was 1,700 feet in length and consisted of a sod surface. (ADAH.)

42 99 - 673 (26)(4-7-34 - 10) (12 - 500) Mobile Airport Constructs

Recognizing the need for an improved facility to accommodate the larger and faster aircraft utilized for commercial passenger and airmail service, city leaders in Mobile initiated the development of a new municipal airport to serve residents. Bates Field, named for City Commissioner Cecil Bates, was located on Cedar Point Road, four miles south of the business district. (ADAH.)

Dedicated in November 1929, the Mobile Municipal Airport, Bates Field consisted of approximately 125 acres of land that included two sod runways, an office building, restrooms, and a telephone. The name of the city was painted in bold letters on the roof of the hangar to orient transient aviators. Operators were assured that the sod runways remained firm even in the wettest weather. (McNeeley Collection, USA Archives.)

By 1939, improvements to Bates Field included a new terminal building, hangar, a concrete ramp, and paved runways. These improvements were funded through the Landing Field Improvement Program of the Works Progress Administration, a federal program created to offset high rates of unemployment created by the Depression. The system of airports in Alabama was significantly expanded because this program. (McNeeley Collection, USA Archives.)

The concrete parking ramp afforded passengers the luxury of boarding aircraft without having to walk through mud or standing water during periods of inclement weather. The airport was equipped for night operations with obstruction and perimeter lighting. A revolving beacon and lighted surface wind indicator were attached to the hangar to assist pilots in locating the airport 'and determining the proper direction to land. (McNeeley Collection, USA Archives.)

Dedicated in May 1939, Hargrove Van de Graaff Field at Tuscaloosa was the second-largest airport in Alabama. Named for former University of Alabama star athlete and original airport sponsor Coleman Hargrove Van de Graaff, the field was described as one of the finest municipal airports in the Southeast. The airport was situated in a natural location that allowed aircraft to be brought in on long, gradual glides for landing. (AFHRA.)

The largest hangar structure of any airfield in the state was located at the new Van de Graaff Field. The new airfield was the third flying field to serve Tuscaloosa. Following the closure of Druid Field in 1925, Maynor Field was dedicated in May 1928. Named for Eldridge W. Maynor, the field was located between the Mobile and Ohio Railroad tracks and the Columbus (Mississippi) Road.

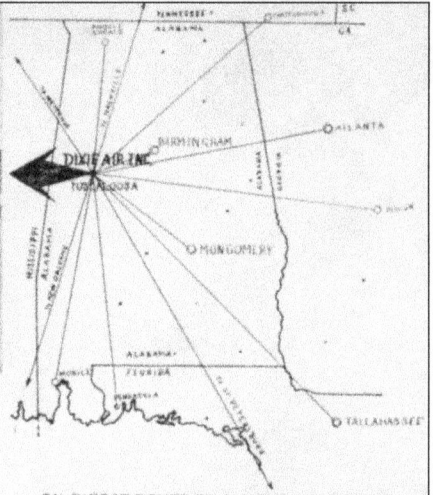

VAN DE GRAAFF FIELD

Dixie Air, Inc.

2 MILES WEST OF *Tuscaloosa, Alabama*

ON DIRECT ROUTE TO ALL THE SOUTHEAST

FINEST FACILITIES IN THE SOUTH

√ HOTEL ROOMS ON THE FIELD
√ CAA WEATHER STATION
√ MODERN DINING FACILITIES
√ CAA APPROVED FLYING SCHOOLS

√ COMPLETE GAS AND OIL SERVICE
√ OVERHAUL AND REPAIR SHOPS
√ LARGE HANGARS
√ AIRCRAFT SALES, RENTALS, REPAIRS

Van de Graaff Field at Tuscaloosa served as an intermediate stop on the Fort Worth–Atlanta air route and provided a direct route to locations in the entire Southeast. Facilities included a Civil Aeronautics Administration–approved weather station and hotel rooms for transient pilots and passengers. Dixie Air provided gas and oil service as well as airplane overhaul and repair shops.

The state of Alabama was a pioneer in the development of aerial air markings. Airports were identified by white lettering painted on the roofs of hangars and other buildings. Arrows pointed along the course of the airway to the next town or airport. Air markings were also painted on the roofs of buildings in the centers of towns and communities to indicate direction to the local airfield.

The introduction of a southern transcontinental airmail route created a new route segment that extended from Montgomery to Atlanta, creating the need for new emergency landing fields. In May 1931, the privately owned Auburn-Opelika Airport was leased to the Bureau of Commerce as Emergency Landing Field No. 33. The airport was officially dedicated in April 1932.

Trained as a military aviator in 1935, Robert Giles Pitts (1910–1992) served as professor of aerospace engineering and director of the School of Aviation at Alabama Polytechnic Institute and subsequently Auburn University for more than 30 years. Pitts was instrumental in the development of the Auburn University Regional Airport that is named in his honor. In 2011, he was inducted into the Alabama Aviation Hall of Fame. (Auburn University Archives.)

In October 1939, Alabama Polytechnic Institute received approval from the Civil Aeronautics Authority to establish a Civilian Pilot Training Program. The program was designed to create a pool of qualified pilots should rapid expansion of military aviation become necessary due to the possibility of global conflict in Europe and Asia. The first pilot training class at API consisted of 20 students. (Auburn University Archives.)

Increased training activity generated by the Civilian Pilot Training Program required improvements at the Auburn-Opelika Airport. These improvements included expansion of the runways, construction of hangar and classrooms facilities, and installation of water and electricity. In addition, the Alabama Polytechnic Institute agreed to assume ownership of the facility as a condition of the Civilian Pilot Training Program. (Auburn University Archives.)

Dedicated in April 1930, the Troy Airport was located two miles south of the city to the west of the paved highway. The field included an east-west runway 2,000 feet in length and a northeast-southwest runway 1,700 feet in length. Established as an auxiliary landing facility for aircraft based at Maxwell Field, Troy Airport was also available for the operation of civilian aircraft.

The Walker County Airport at Jasper was one of 42 airport projects in Alabama completed under the direction of the Civil Works Administration and the Works Progress Administration. These programs were administered by the new Alabama aviation commission, established in 1935. Theodore Swann of Birmingham served as the first chairman of the commission. In October 1945, the aviation commission was reorganized into the Alabama Department of Aeronautics.

In February 1939, Eastern Air Lines was awarded a certificate of public conveyance and necessity by the Civil Aeronautics Authority for Commercial Airmail Route 40 connecting the cities of Miami and Memphis. Florence, Sheffield, and Tuscumbia dispatched mail through the airport at Muscle Shoals. The cities of Birmingham, Montgomery, and Dothan were also served on the route. (Muscle Shoals Airport.)

In June 1931, the Huntsville Junior Chamber of Commerce hosted a dedication ceremony for the Mayfair Aviation Field, the city's first airport. More than 4,000 people attended the weekend event that included airplane races, parachute demonstrations, and aerial aerobatics. L.G. Mason (shown in front of airplane) of Montgomery won the 125-horsepower class airplane race during the dedication ceremonies. (Huntsville-Madison County Public Library.)

In 1940, Madison County and the City of Huntsville purchased a 720-acre parcel of land for construction of a new airport. The Work Projects Administration provided funding for the construction of three sod runways and a small wood-frame building. The original administration building shown in this image was utilized as an air cargo building by Eastern Air Lines following construction of a new terminal building in 1951. (Huntsville-Madison County Public Library.)

In November 1944, Pennsylvania Central Airlines inaugurated service to Huntsville on Commercial Airmail Route 55 from New York to Birmingham. The 21-passenger Douglas DC-3 used on the inaugural service was christened *The City of Huntsville* by resident Susie Spragins. Assisted by Rudolph Scott, a local minister, the aircraft was christened with water from Big Spring, the original water source for the city of Huntsville. (Huntsville-Madison County Public Library.)

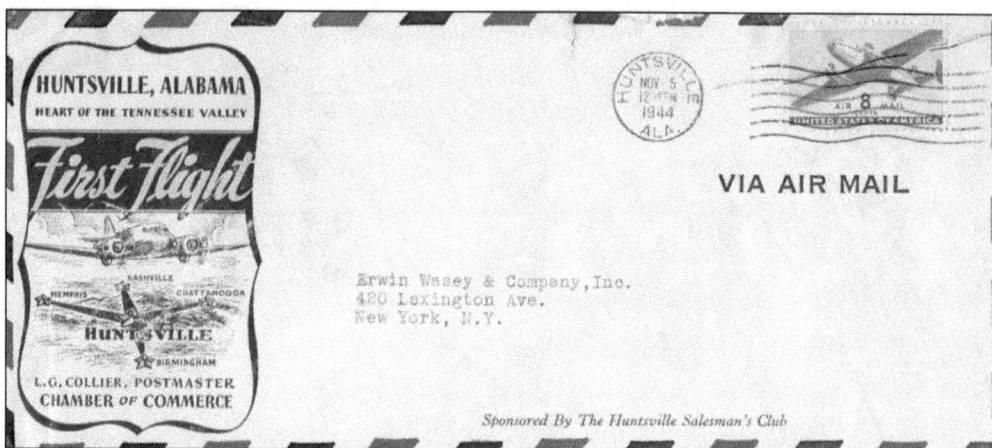

Inauguration of service to Huntsville by Pennsylvania Central Airlines was commemorated by a first-day-of-service postal cover. The cachet graphically displays cities connected to Huntsville by aerial mail and passenger service. Service to Memphis and Nashville was provided by Eastern Air Lines. New service to Birmingham and Chattanooga was provided by Pennsylvania Central Airlines.

Headquartered at Washington, DC, Pennsylvania Central Airlines was known as the "Capital Airline." In 1948, the name of the company was changed to Capital Airlines to reflect its expanding route system. The aircraft of Capital Airlines were identified as Capitaliners. In this image, passengers deplane from a Capital Airlines aircraft on the ramp of the Huntsville Municipal Airport. (Huntsville-Madison County Public Library.)

Subsequent improvements to the Huntsville Municipal Airport included construction of a new terminal building, control tower, and paved runways to accommodate larger commercial aircraft. After the Second World War, additional service was operated by Southern Airways and Waterman Airlines. Following the opening of the new Huntsville-Madison County Jetport in October 1967, the former municipal airport was closed and subsequently developed as a recreational facility. (Huntsville-Madison County Public Library.)

A Douglas DC-3 of the Great Silver Fleet of Eastern Air Lines departs for Nashville on Commercial Airmail Route 40, with service from Birmingham to New York, inaugurated in January 1946. Huntsville became an intermediate stop between the cities of Nashville and Birmingham. During the pre-security days of air travel, visitors parked their cars adjacent to the ramp area to watch flights depart. (Huntsville-Madison County Public Library.)

In January 1934, the City of Dothan purchased 100 acres of land located three miles west of the city for construction of a municipal airport. Initially, Eastern Air Lines would agree to use the airport only during conditions of favorable winds because of the restricted length of the runways. Following completion of a project to lengthen the runways and construct a terminal building, daily passenger and mail service was inaugurated in August 1940 on Commercial Airmail Route 40 from Miami to Memphis. The facility was closed in February 1966 as the city began utilizing the former Napier Army Air Field as a municipal airport. Westgate Park was developed on the site of the former municipal airport. (Both, AFHRA.)

Four

ARSENAL OF DEMOCRACY

During the Second World War, 10 installations located in the state of Alabama provided primary, basic, advanced, or specialized four-engine flight training to military aviation cadets. Maxwell Field served as headquarters of the Eastern Flying Training Command supervising all pilot, navigator, and bombardier training in the eastern United States. Brookley Field in Mobile provided logistical and maintenance support for training bases throughout the Southeast.

Greetings from MAXWELL FIELD, Alabama

Maxwell Field in Montgomery was originally established as an aviation maintenance depot to support flight training operations in the Southeast during the First World War. In subsequent years, Maxwell Field would become headquarters of the Army Air Corps Tactical School and Army Air Corps and Army Air Forces Training Commands. From 1940 to 1945, more than 60,000 military aviation cadets were trained as pilots under the supervision of Maxwell Field.

Military personnel and aircraft have always provided support to the people of Alabama during periods of national emergency and natural disaster. In March 1929, heavy rains caused extensive flooding of towns and communities across south Alabama. During the five-day relief effort, aircraft from Maxwell Field were used on 281 missions and dropped more than 50 tons of supplies to people stranded in flooded areas.

Constructed as an academic facility for the Air Corps Tactical School, Austin Hall served as headquarters of the Southeast Air Corps Training Center during the Second World War. One of only three such training centers, the Southeast Center had jurisdiction over all pilot, bombardier, and navigator training programs in the southeastern United States. The building was named for Lt. Charles Austin, a former instructor at the tactical school. (AFHRA.)

In the spring of 1931, the Army Air Corps Tactical School was relocated to Maxwell Field from Langley Field, Virginia. Graduates of the school provided the leadership for development of the airpower doctrine during the Second World War. Of 320 Army Air Forces general officers on active duty at the conclusion of the Second World War, 261 were graduates of the AAC Tactical School at Maxwell Field. (AFHRA.)

The AAC Tactical School coat of arms shown on the fuselage of this Consolidated PB-2A pursuit aircraft includes the motto "Proficimus More Irrententi." This Latin phrase is translated as "We Make Progress by Custom Unhindered." The green and blue colors of the shield represent the earth and sky. The four bolts of lightning represent the branches of the Air Corps as they existed in 1929—Pursuit, Bombardment, Attack, and Observation. The flaming lamp of knowledge is winged to symbolize the guiding light of the United States Army Air Corps. The insignia was originally designed in 1929 for the Air Corps Tactical School at Langley Field, Virginia.

By 1932, the transition of the former aviation engine and repair depot into a modern military aviation facility had become apparent. In addition to the expansion of the flying field, new hangar facilities were constructed in proximity to the former maintenance structures erected during the First World War. Base facilities were initially expanded to accommodate the relocation of the Army Air Corps Tactical School from Langley Field, Virginia.

This 1941 image of Maxwell Field illustrates the continued growth that occurred due to the expansion of military aviation prior to the Second World War. The most evident change is the addition of hard-surface runways constructed to accommodate heavy-bombardment and transport aircraft. (AFHRA.)

In September 1941, the Army Air Corps Replacement Center was established at Maxwell Field. The first of its kind, the center provided indoctrination in military protocol and other academic subjects prior to enrollment in specialized training. Cadets were required to complete 60 hours of training in activities designed to achieve peak physical conditioning. In 1942, the center was redesignated the Army Air Forces Preflight School. (AFHRA.)

The base operations building initially served as the headquarters of Maxwell Field. The second story of the building included offices of the executive officer, the clerical staff, and weather bureau. The facility also included an area for repairing, repacking, and testing parachutes. The building was constructed on the approximate site of the hangar of the first civilian flying school established by the Wright brothers.

Because of the rapid expansion of the military flight training program, a shortage of flight instructors threatened to severely limit the number of cadets trained in the basic and advanced flying schools. To alleviate this shortage, the Central Instructors School was established at Maxwell Field in 1942. In addition to training new instructors, the staff developed more efficient and standardized methods of flight instruction.

To accommodate increasing numbers of students in the advanced flying schools at Maxwell Field, link trainers were temporarily housed in a hangar on the base. The link trainers were utilized to train cadets in the operation of an aircraft without outside visual references. Instructors were stationed at a desk behind each trainer to monitor the student's control of the device and proficiency in cross-country navigation.

In June 1941, the Army Air Corps became a subordinate element of the Army Air Forces. In July 1943, flight training at Maxwell Field entered a new era with the establishment of the Army Air Forces Pilot School, Specialized Four-engine. The runways at Maxwell Field were reinforced and extended to 5,500 feet in length to accommodate the four-engine Consolidated B-24 Liberator heavy-bombardment aircraft. (AFHRA.)

In 1945, pilot training for the B-24 Liberator heavy-bombardment aircraft was transferred to Courtland Army Air Field as Maxwell Field began transition training for pilots and flight engineers on the Boeing B-29 Superfortress. To accommodate the giant aircraft, the runways at Maxwell were extended to 7,000 feet in length. The War Department also authorized the construction of the largest maintenance hangar to be built on the field. (AFHRA.)

The War Department selected the Southeast Training Center to train 8,000 members of the Royal Air Force of Great Britain for combat service. The first 750 British cadets arrived at Maxwell Field in the fall of 1941. After completing training at the Basic Flying Training School at Gunter Field, the British cadets were assigned to advanced schools at Maxwell Field, Napier Field in Dothan, or Craig Field in Selma.

"This quiet corner of a distant foreign field shall forever be England." That distant foreign field is the Oakwood Cemetery in Montgomery. Flight Lt. George W. Nickerson is one of 78 members of the Royal Air Force who lost their lives while attending training in Montgomery during the Second World War. In an adjacent section of the cemetery, 20 French airmen were laid to rest.

Trainers in Flight
COURTLAND ARMY AIR BASE
COURTLAND, ALABAMA

Photo courtesy
U. S. Army Air Forces

Activated in December 1942, the basic flying school at Courtland Army Air Field was located southwest of the town of Courtland in Lawrence County. The area selected for the site had previously consisted of large cotton plantations that included several small tenant houses. The 4,000 military and civilian personnel assigned to the base was more than five times the population of Courtland.

Construction of the 2,236-acre Courtland Army Air Field was completed in six months. The runway complex was designed in an interlocking, wraparound pattern that allowed uninterrupted flight training by eliminating lost time adjusting to changes in wind direction. Consisting of 18 inches of steel-reinforced concrete designed to accommodate the heaviest bomber aircraft, the runways were the strongest of any of the 35 air bases of the Eastern Training Center. (AFHRA.)

Col. C.P. West, commanding officer of the basic flying school at Courtland, piloted the first aircraft to land at the new facility. Colonel West then participated in a brief ceremony in which the flag of the post was raised for the first time. During the ceremony, West complimented civilian officials and local residents for their splendid cooperation during construction of the field. (AFHRA.)

In February 1943, the first group of cadets arrived at the basic flying school at Courtland. The basic flying course followed completion of primary training. After graduation from the basic school, the cadets would progress to advanced training in single-engine pursuit or large four-engine bombardment aircraft. Not every cadet arriving at Courtland would complete the training program as washout rates limited the number of graduates. (AFHRA.)

Construction of more than 300 buildings on the base, including the large maintenance hangar, required more than five million board feet of lumber. The Army Corps of Engineers estimated that the total amount of concrete in the roads, aprons, and runways of the base was equivalent to that required to build a road 22 feet in width from Decatur to Muscle Shoals, a distance of 50 miles. (AFHRA.)

Personnel in the air traffic control tower maintained an orderly flow of air traffic during hundreds of daily air operations. The high level of flight-training activity combined with inexperienced cadets occasionally resulted in tragic accidents. On the night of December 1, 1943, eight student pilots were fatally injured in crashes after becoming disoriented and exhausting their fuel supply on a night cross-country training flight. (Town of Courtland.)

More than 200 Vultee BT-13 training aircraft were based at Courtland. Known as the Vultee Vibrator because of the noise and vibration produced by the engine, the aircraft was used for cadet training in formation as well as cross-country and night flying. To minimize congestion at the main airfield, five auxiliary landing fields were utilized—Danville, Trinity, Bay, Leighton, and the Muscle Shoals Municipal Airport. (Town of Courtland.)

For aviation cadets recently graduated from primary training, the more advanced aircraft and course of instruction in the basic phase proved extremely challenging. To minimize time aloft spent learning the basics of aircraft radio operations, technicians developed a ground-based trainer. The trainer duplicated the radio system of the BT-13 training aircraft, allowing cadets to develop proficiency in radio procedures without leaving the ground.

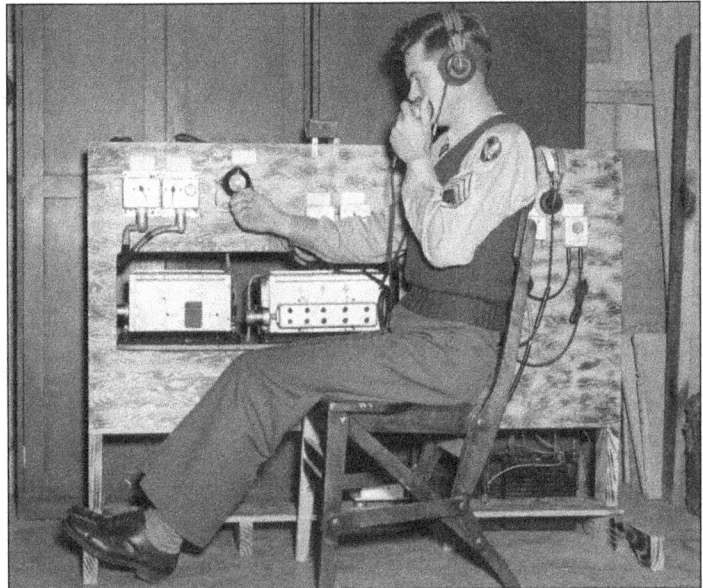

High Over Alabama

In September 1944, the mission of the Courtland Army Air Field was changed from a basic flying school to a Specialized Four-Engine Transition School. The aircraft utilized was the Consolidated B-24 Liberator heavy-bombardment aircraft. The first B-24 aircraft arrived at Courtland in August 1944 for use in familiarization by maintenance personnel. The parking apron on the airfield could accommodate 62 B-24 aircraft.

Routine maintenance of the B-24 heavy-bombardment aircraft was performed in 13 maintenance docks. Each dock would provide partial covering of one B-24. The cost to construct one maintenance dock was $2,000, as compared to $66,000 for a fully enclosed hangar. Because the base was classified as Emergency Temporary Construction, only one permanent maintenance hangar was constructed to perform extensive aircraft maintenance. (AFHRA.)

11—Advanced Trainer during High Altitude Flight, Craig Field, Selma, Ala.

PHOTO BY SOUTHEAST AIR CORPS TRAINING CENTER

Craig Army Air Base at Selma was named for Bruce Kilpatrick Craig (1924–1941). A native of Selma, Craig formed a glider club and received his pilot's certificate while in high school. Prior to receiving a commission in the Army Air Corps, Craig was fatally injured during flight testing of a Consolidated B-24 heavy-bombardment aircraft. He was posthumously awarded a commission as a second lieutenant.

Upon completion, Craig Army Air Field was one of the largest aviation facilities in the United States. The post was composed of approximately 2,000 acres of almost level ground. One-half of the total acreage was utilized as a flying field two miles in length and one mile in width. The first of 150 aircraft arrived at Craig Field the first week of April 1941, with flight operations beginning the following week. (AFHRA.)

The site selected for the new Air Corps advanced flying school was an area of large cotton fields located along US Highway 80, six miles east of Selma. In August 1940, Mayor Lucien Burns turned the first spade of dirt during a ceremony dedicating the site of what would become one of the largest military flying schools in the United States. (AFHRA.)

From 1941 until 1945, more than 6,000 military aviators were trained at Craig Field, including 2,500 British, French, and Dutch cadets. Training classes included map reading, meteorology, radio procedures, and flight by reference to aircraft instruments. Upon activation, Craig Army Air Field became the first specialized single-engine training school in the Southeast. (AFHRA.)

To accommodate more than 3,000 military support staff and cadets assigned to the school and service squadrons, Craig Field became a miniature town with airfield, administration, and classroom buildings. The post also had a chapel, repair shops, theater, weather offices, and a base exchange. Because of a shortage of housing, a 100-unit duplex-apartment complex was constructed in proximity to the base. (AFHRA.)

The flight training program was initiated before the infrastructure of the base was completed. Initially, aircraft servicing and maintenance was performed in a group of tents located along the flight line. With completion of the aircraft parking apron, a new maintenance hangar was constructed to create a protected area for maintenance personnel to service more than 150 training aircraft operating on the field. (AFHRA.)

The training program at Craig Field was the first in the Southeast to provide instruction in pursuit and gunnery in addition to the prescribed advance flying course. Other specialized areas of training included high-altitude flights, formation flying, and air-to-air combat maneuvers. The program was also one of the first to include transition training to pursuit aircraft, a 10-hour course of instruction in the Curtiss P-40 single-seat fighter. (AFHRA.)

By September 1941, Craig Field contained more than 175 acres of streets, runways, taxiways, and parking areas. The advanced flying school was the largest flying field, measured by acreage, in the United States. Five auxiliary landing fields were assigned to the flight operations at Craig Field— the Selma Municipal Airport, Furniss Field at Cahawba, Henderson Field at Millers Ferry, Mollette Field at Orville, and Autaugaville. (AFHRA.)

CLASS SE-41-G
CRAIG FIELD, SELMA, ALABAMA

Graduation exercises for the initial classes of cadets were held at the Wilby Theater in downtown Selma. On what was proclaimed "Flying Cadet Day," citizens of Selma were encouraged to display American flags and other suitable patriotic symbols in observance of the occasion. As weather permitted, subsequent graduation ceremonies were held on the post in a grove of pecan trees that formed a small park between the academic buildings. (AFHRA.)

Celebrities and other notable individuals often visited military bases across the country as part of United Service Organizations events or simply to increase the morale of military personnel. Comedian Bob Hope organized numerous domestic and overseas performances of movie stars, musicians, and athletes. Boxing legend Jack Dempsey, shown sitting in the cockpit of a North American AT-6 advanced trainer, visited troops at basic and advanced flying schools in Alabama.

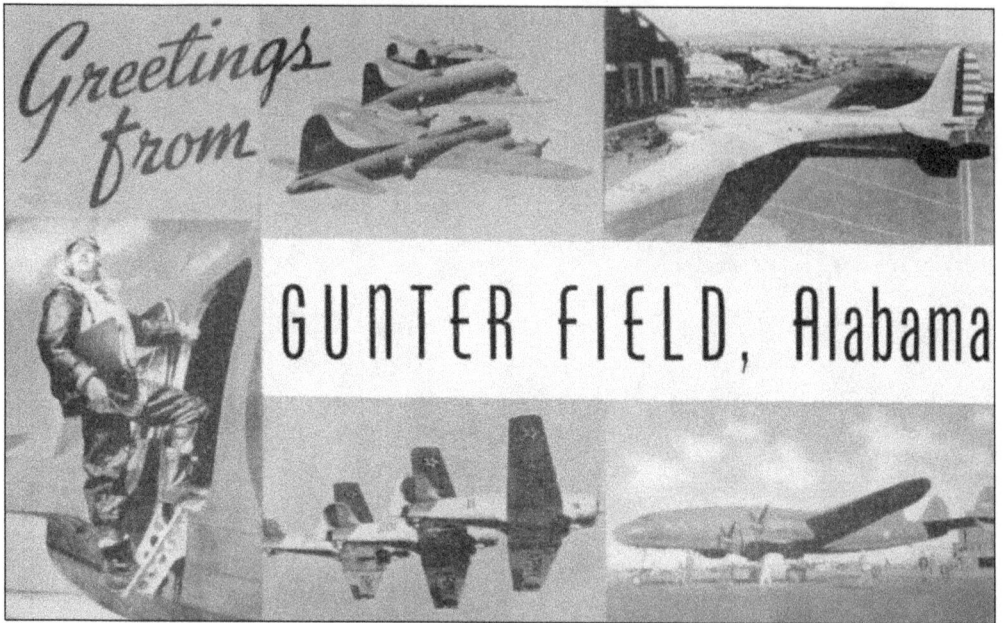

Greetings from GUNTER FIELD, Alabama

Activated in August 1940, the airfield of the Army Air Corps' basic flying school in Montgomery was renamed for William Adams Gunter, former mayor of Montgomery and proponent of aviation in the capital city. Gunter Field became the first United States Army Air Corps basic flying school in the Southeast. From 1940 until 1945, twelve thousand American, British, and French cadets completed basic flight training at Gunter Field.

The site selected for the Basic Flying Training School was the Montgomery Municipal Airport. Military planners determined that military flying schools could be established and activated in shorter periods of time by leasing existing municipal airports. The lease agreement for Gunter Field included a provision that allowed continued operations of daily Eastern Air Lines service to the field as long as the commercial flights did not interfere with military flight training. (AFHRA.)

The construction of Gunter Field represented a link between the Great Depression and the Second World War. The former municipal airport was converted into a military training facility by the Work Projects Administration. The base was not only used for national defense but also created a huge economic benefit to Montgomery. The transformation of Gunter Field became a model for construction of military bases and airfields throughout the United States. (AFHRA.)

A native of Eclectic, Alabama, Aubrey Hornsby (1895–1982) served as an aerial observer during the First World War. After the war, Hornsby served as a flight instructor at Kelly Field, Texas. He participated in the sinking of the German battleship *Ostfriesland* in a demonstration of airpower organized by Gen. Billy Mitchell. In 1940, Hornsby became the first commanding officer of the Basic Flying Training School at Gunter Field. (AFHRA.)

The basic training aircraft utilized at Gunter Field was the Vultee BT-13 Valiant. Powered by a 450-horsepower Pratt & Whitney Wasp engine, the aircraft had a top speed of 180 miles per hour. The BT-13 had a tandem seating arrangement with the flight instructor normally occupying the rear seat. Cadets and instructors often flew with the canopy open to alleviate the summertime heat.

During 1944, more than 400 training aircraft were utilized at Gunter Field. When these were combined with training at Maxwell Field, the skies over Montgomery were described as having the densest air traffic in the world. Six auxiliary landing fields were used to relieve air traffic congestion at Gunter Field—Shorter, Mount Meigs, McLenmore, Deatsville, Elmore at Wetumpka, and Taylor Field, the former First World War flying field at Pike Road.

The combination of intensive training schedules and inexperienced pilots occasionally created disastrous situations. In May 1942, a group of aircraft departed Gunter Field on a night cross-country training flight. On the return leg, the flight encountered deteriorating weather conditions that caused the student pilots to become disoriented. As their fuel supplies were exhausted, 12 of the 35 aircraft crashed while attempting to land. Seven of the cadets were fatally injured.

In addition to normal approaches and landings, cadets received instruction in landing over an obstacle. In what was known as the hurdle stage of landing practice, the student aviators were evaluated on their ability to fly over a simulated obstacle and land within a specified area. Cadets were required to complete flight tests at various stages of training. Failure to achieve satisfactory performance resulted in the cadet being washed out of the program.

Although Eastern Air Lines was initially allowed to continue passenger and mail service to the facility, increased training activity at Gunter Field resulted in the airspace becoming too congested for continued civilian operations. In March 1941, a 508-acre parcel of land located west of Montgomery on the Selma Highway was selected as the site of a new municipal airport. Upon completion, the new facility consisted of three paved runways and the most modern airport lighting system available. In November 1942, the new municipal airport was leased by the War Department for use as an auxiliary landing field for aircraft operating from Gunter Field. Dedicated in July 1943, the facility was named for Montgomery native Clarence Moore Dannelly Jr., who lost his life in an aviation accident at the Pensacola Naval Air Station. (Above, AFHRA; left, Montgomery Airport Authority.)

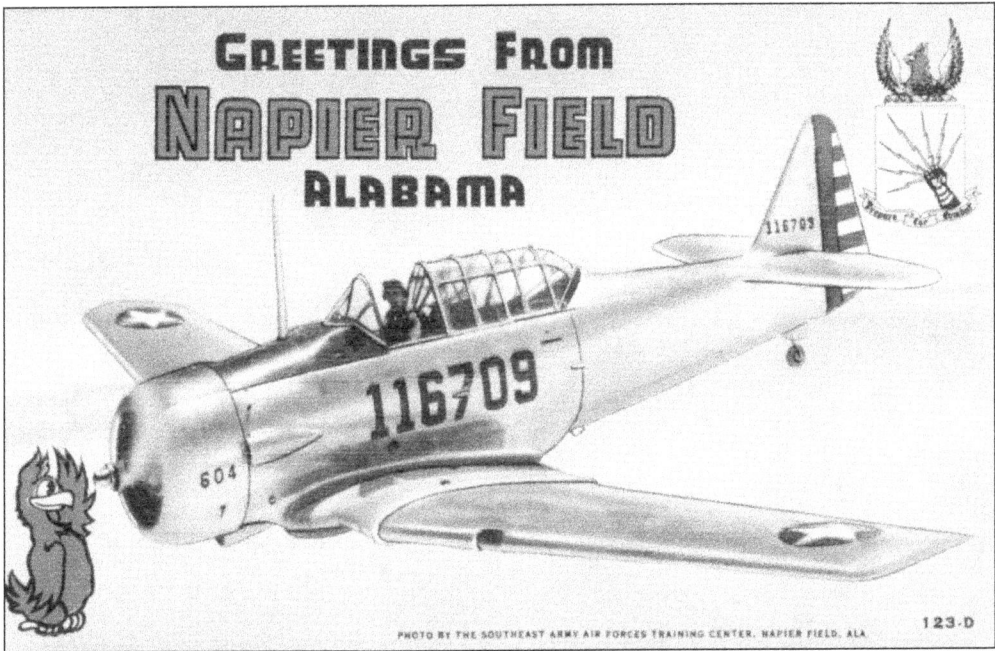

PHOTO BY THE SOUTHEAST ARMY AIR FORCES TRAINING CENTER, NAPIER FIELD, ALA. 123-D

In April 1941, Sen. Lister Hill announced that a 1,500-acre parcel of property located on the Ozark-Dothan Highway near the Grimes community had been selected as the site of a new Army Air Corps advanced flying school. The facility would be named for Maj. Edward L. Napier of Union Springs, Alabama. Napier lost his life during flight testing of a captured German Fokker D-VII aircraft at McCook Field, Ohio.

Upon completion, Napier Field consisted of four paved runways, each 4,500 feet in length and 300 feet in width. The facility also included a concrete ramp and warm-up apron 5,200 feet in length and 380 feet in width. The post included two large maintenance hangars and 208 buildings. (AFHRA.)

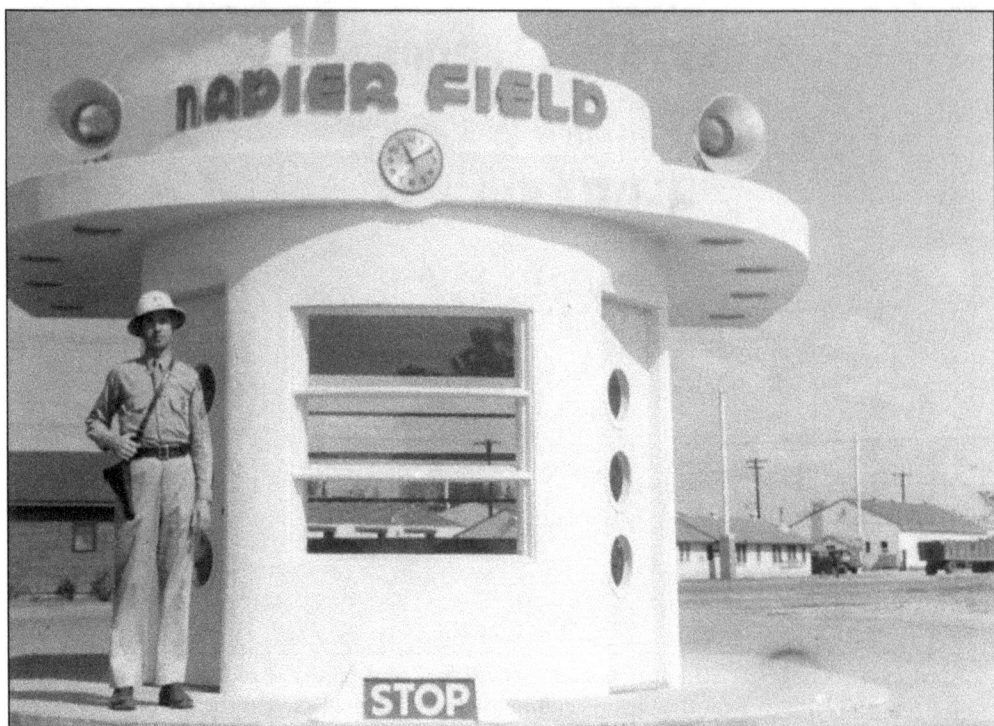

In October 1941, slightly more than three months since construction of Napier Field began, the first airplane landed on one of the newly completed runways. The aircraft, based at Craig Field in Selma, carried payroll for the troop detachment stationed on the post. Only six months elapsed from the first shovel of dirt being turned until the initiation of full-scale flight operations in December 1941. (AFHRA.)

Personnel of the Napier Field Post Headquarters supervised 188 officers, 352 flying cadets, 2,015 enlisted men, and 15 nurses. From December 1941 until June 1945, instructors at Napier Field trained American, British, and French military aviators for service overseas. In addition to male aviators, the Women Air Force Service Pilots served a vital role at the field, including aircraft test flights and daily inspections of auxiliary landing fields. (AFHRA.)

Military controllers directed aircraft operations from the air traffic control tower. From the beginning of flight training in December 1941 until the last class of cadets was graduated in 1945, the role of Napier Field expanded from an Advanced Pilot Training School to providing additional pilot transition training in frontline fighter aircraft such as the Republic P-47 Thunderbolt and the North American P-51 Mustang. (AFHRA.)

More than 6,000 aviators graduated from Napier Field before being deployed to overseas assignments, including 1,268 pilots of the Royal Air Force. Many of these aviators had previously completed primary and basic flight training in Alabama. Alabama native Henry Condon, a graduate of advanced training at Napier, would be credited with five enemy aircraft destroyed in the Pacific theater of combat. (AFHRA.)

Based on a ratio that allowed a maximum utilization of aircraft, 142 North American AT-6 and 14 Vultee BT-13 training aircraft were assigned to the advanced flying school at Napier Field. To accommodate the high rate of utilization of these aircraft, five auxiliary landing fields were utilized—Minor Field at Wicksburg, Benoit Field at Headland, Hyman at Columbia, Ozark Army Air Base, and the Dothan Municipal Airport. (AFHRA.)

In preparation for aerial gunnery practice, cadets at Napier Field received training in a ground-based platform. On a remote section of the base, mounds of dirt with targets attached were constructed for the firing of machine guns and small cannons. After demonstrating proficiency in the ground-based firing platform, cadets proceeded to Eglin Field in Florida for actual aerial gunnery practice using ground-based targets. (AFHRA.)

The cadet mess hall and barracks at Napier Field were temporary construction buildings. These simple wood-frame structures consisted of inexpensive, prefabricated materials that could be built in an assembly-line manner. Materials were bundled into construction packages that would meet the needs of 125 men and included barracks, mess halls, and recreation buildings. (AFHRA.)

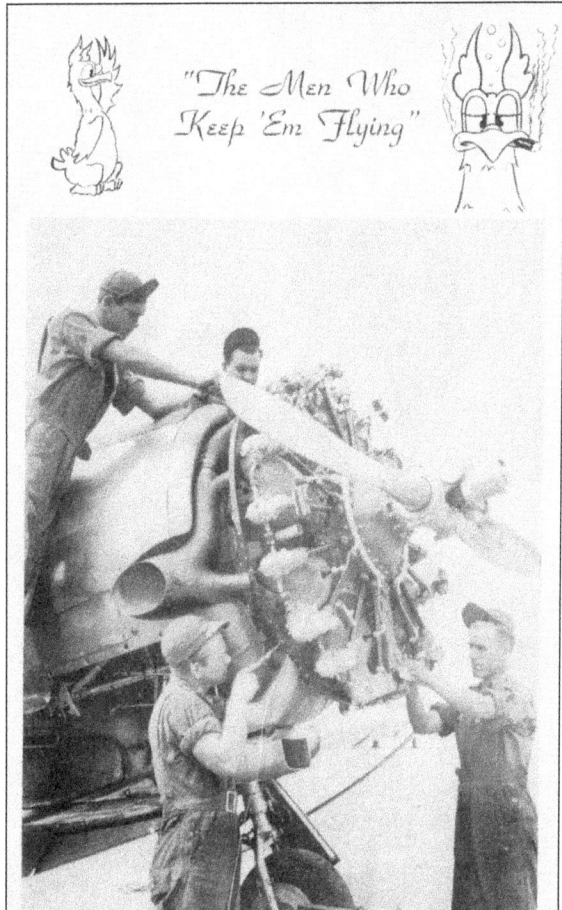

The "Men Who Keep 'Em Flying" were the aircraft mechanics and technicians of Napier Field. Because of high utilization rates and frequent minor accidents involving inexperienced student aviators, maintenance personnel worked in eight-hour shifts around the clock to keep aircraft in the air. The bird in the image is Gruffy, mascot of Napier Field. Gruffy was a legendary bird of outstanding characteristics and bravery.

In March 1941, First Lady Eleanor Roosevelt traveled to Tuskegee to attend a meeting of the Julius Rosenwald Foundation. Against the objections of her protective detail, Roosevelt asked Charles Alfred Anderson (1907–1996) to take her on a flight over the Tuskegee campus in a Piper J-3 Cub. The historic flight originated from Kennedy Field, located on the Union Springs Highway south of Tuskegee. (AFHRA.)

The Waco UPF-7 was the first training aircraft utilized in the Civilian Pilot Training Program at Tuskegee Institute. Initially, students completed academic training provided by Robert Pitts and Bloomfield Cornell of the Alabama Polytechnic Institute. The flight training program was contracted with Alabama Air Service in Montgomery. In February 1940, flight training was moved to Airport No. 1, Kennedy Field.

Considered the "Father of African American Aviation," Charles Alfred "Chief" Anderson was instrumental in the formation and development of the Civilian Pilot Training Program at Tuskegee Institute. Anderson would later serve as chief instructor of the 66th Army Air Forces Flight Training Detachment at Moton Field. As chief instructor, Anderson was directly involved in training the 992 African American aviators who would become the Tuskegee Airmen. (AFHRA.)

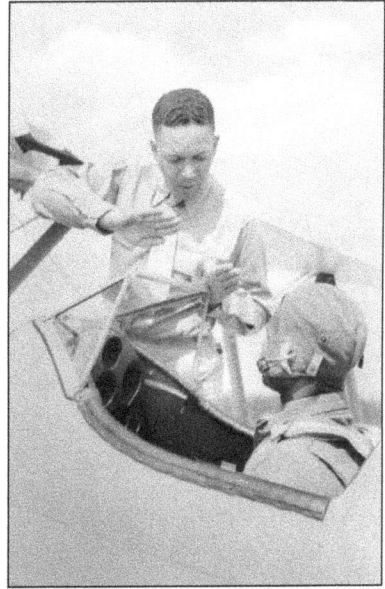

Following the Second World War, Anderson continued to train civilian pilots at Moton Field in Tuskegee and promote aviation and aeronautical education among African Americans for more than 50 years. In an aviation career that encompassed seven decades, "Chief" Anderson amassed more than 50,000 hours of experience in the air. In 1991, Anderson was inducted into the Alabama Aviation Hall of Fame.

Born in Tuskegee, Forrest McKenzie Shelton Jr. (1919–1999) was instrumental in the establishment of the first flying field in Tuskegee. Shelton (left) served as one of the first flight instructors of the 66th Army Air Forces Flight Training Detachment at Moton Field, providing flight instruction to the first classes of Tuskegee Airmen. In 2003, Shelton was inducted into the Alabama Aviation Hall of Fame. (AAHOF.)

Noel Francis Parrish (1909–1987) graduated from Cullman High School in Alabama. A career Army Air Corps pilot, Parrish developed an early interest in promoting the involvement of African Americans in military aviation. Lieutenant Colonel Parrish assumed command of Tuskegee Army Air Field in December 1942 and oversaw the training of African American aviators through the end of the Second World War. Parrish retired from the military as a brigadier general. (AFHRA.)

Consisting of 1,650 acres of land located seven miles northwest of Tuskegee, the Tuskegee Army Air Field was the only facility in the United States established to provide basic and advanced flight training for African Americans pilots of the armed forces. Although construction of the field had not been completed, pilot training programs began in November 1941. The post consisted of four paved runways and 225 buildings. (AFHRA.)

Aircraft at the Tuskegee Army Air Field were maintained in three maintenance hangars. The structures are identified by the checkerboard pattern painted on the roofs. Following deactivation of the base, the War Assets Administration transferred ownership of the hangar structures to communities for use at local municipal airports. The structures were disassembled and transported to airports at Montgomery, Troy, and Clanton. (AFHRA.)

The Post Headquarters of the Tuskegee Army Air Field supervised all operations on the base. The basic and advanced flight training programs were staffed by flight and ground instructors, air traffic controllers, maintenance technicians, meteorologists, and medical and other support personnel. Activated in July 1941, the first troop contingent arrived at the base in October 1941. By 1942, more than 3,000 military and civilian personnel were stationed at the field. (AFHRA.)

The Base Operations Building (left) controlled every aspect of flight operations at Tuskegee Army Air Field. Meteorologists assigned to the base monitored weather conditions using recording instruments installed on the roof of the building. The radio frequency of the base control tower is painted on the roof of an adjacent building to alert pilots of transient aircraft. (AFHRA.)

In July 1941, thirteen cadets led by student officer Capt. Benjamin O. Davis reported to the Tuskegee Institute to begin training as the first class of African American pilot candidates in the United States military. In March 1942, the first class of aviation cadets to graduate from the Tuskegee Army Air Field became the nation's first African American military pilots. Pictured above are, from left to right, George "Spanky" Roberts, Benjamin O. Davis, Charles DeBow Jr., Instructor Pilot R.M. Long, Mac Ross, and Lemuel Custis brief a training flight. Pictured below, gathered around the cockpit of a Vultee BT-13A, are, from left to right, DeBow, Custis, Ross, Davis (seated in aircraft), Roberts, and Long. Davis would become the first African American to achieve the rank of general officer in the United States Air Force. (Both, AFHRA.)

Cadets receive a training mission briefing before takeoff from the Tuskegee Army Air Field. Following the briefing, the instructor accompanied each student on an individual training flight. Because of the high level of training activity, cadets utilized auxiliary landing fields at Shorter and Tallassee for takeoff and landing practice. (AFHRA.)

A cadet at the Tuskegee Army Air Field prepares for a training session in the Link Trainer. Before the field was deactivated in June 1946, nine hundred ninety-two pilot cadets had completed basic and advanced training at the facility. Graduates of the flight and ground support programs at Tuskegee would serve with honor and distinction in theaters of combat in North Africa, Sicily, and the Italian mainland. (AFHRA.)

Established during the rapid expansion of naval aviation during the Second World War, Naval Auxiliary Air Station Barin Field was constructed on the site of the Foley Municipal Airport. Commissioned in December 1942, the field was named for Lt. Louis Theodore Barin, naval aviator No. 56, who lost his life in June 1920 in an aviation accident in San Diego, California. (National Museum of Naval Aviation.)

Barin Field consisted of two separate operational areas—the East and West Fields, which were identified by their relative direction from the hangar complex. Each field consisted of four runways that permitted takeoffs and landings in eight directions, eliminating the necessity of operating in a crosswind of more than 22 degrees. The field operated under the jurisdiction of the Pensacola Naval Air Training Center. (National Museum of Naval Aviation.)

The only naval auxiliary air station in Alabama, Barin Field was established to provide cadets with advanced training in fighter and other carrier-based aircraft. Established in rural Baldwin County, the field was located in proximity to Naval Air Station Pensacola yet isolated enough to not interfere with flight training operations at other outlying fields of the naval air center. (National Museum of Naval Aviation.)

By April 1944, more than 400 training aircraft were based at Barin Field. North American SNJ aircraft were utilized to provide training in aerial gunnery, bombing, aircraft carrier landing qualification, and night flying. During the first 24 months of operation, 5,795 naval aviators completed training at Barin Field. (M.L. Shettle Jr.)

Runway markings at Barin Field were identical to those of an aircraft carrier. Student aviators, assisted by a landing signal officer, would make simulated carrier landings (traps) until becoming proficient in the maneuver. These simulated carrier landings were the most hazardous aspect of the training program. The student would then advance to the next stage of qualification, solo landings on an aircraft carrier. (National Museum of Naval Aviation.)

Although a cadet completed a rigorous training program before attempting his first landing on an aircraft carrier, accidents were inevitable. The combination of inexperienced aviators, a confined landing area, and the continuous movement of the ship made this stage of training a formidable task. The first landings on the aircraft carrier were made without the assistance of a flight instructor. (National Museum of Naval Aviation.)

Army Air Base
BROOKLEY FIELD, ALABAMA

In July 1939, the War Department designated Mobile as the site of a new military aviation maintenance and supply depot. Initially designated the Southeast Air Deport, the facility was established to provide maintenance and support for military aviation operations in five southeastern states and the Caribbean area, including Puerto Rico. During the Second World War, the depot operated under the jurisdiction of the United States Army Air Forces Materiel Command.

The site selected for the depot was located four miles south of the city, adjacent to Mobile Bay. The War Department originally acquired 1,362 acres of land that included the municipal airport. This 1941 aerial image shows construction of the repair hangars, shops, and airfield nearing completion. The racetrack used for early aerial exhibitions in Mobile is visible in the lower left corner of the image. (Brookley Collection, USA Archives.)

During 1943, the Mobile Air Service Command headquartered at Brookley Field supplied 31 airfields in five states—Alabama, Tennessee, Mississippi, Louisiana, and the section of Florida located west of the Apalachicola River. The infrastructure of Brookley Field consisted of 4.1 million square feet of floor space under roof in maintenance shops, hangars, warehouses, and administrative offices. (Brookley Collection, USA Archives.)

During 1943, more than 17,000 civilian workers were employed at Brookley Field. Because of the wartime shortage of labor, approximately one-half of the workers employed were women, making the facility the largest employer of women in Alabama. In addition to performing clerical tasks, women were trained as mechanics to repair and service airframe and engine components and as inspectors to approve aircraft modifications. (USA Archives.)

The Mobile Air Service Command was divided into two general categories: maintenance and supply. By 1943, the base consisted of 512 buildings, 26 miles of roads, and approximately 9 miles of railroad track to facilitate receiving, storing, and shipping of parts and supplies. Brookley Field was unique in being the only military aviation installation in the United States served directly by four modes of transportation and a deepwater port.

The Mobile Air Service Command performed routine maintenance, modifications, and repairs on a variety of military aircraft that included the B-24 Liberator and B-29 Superfortress heavy-bombardment aircraft. During the final six months of 1943, approximately 19,000 operations that included maintenance, test, and international cargo flights, were conducted from the Brookley Complex.

This aerial image of Brookley Field illustrates the massive expanse of the facility. The housing area in the upper right corner of the image was previously the location of Legion Field, the first municipal airport to serve the city of Mobile. The original Bates Field was located in the triangular area formed the by the three perimeter runways. (University of Alabama Map Library.)

Construction of Brookley Field on the site of the original Mobile Municipal Airport required the city commission to seek a location to construct a new municipal airport. An 816-acre parcel of property located 12 miles northwest of the city at the intersection of Government Street and Dawes Road became the new Bates Field. The new facility consisted of three paved runways and a municipal hangar for the storage of aircraft. (Pamela Englund.)

Birmingham Army Air Base
BIRMINGHAM, ALABAMA
Official Photograph, U.S. Army Air Forces

In August 1939, the War Department compiled a list of civil airports that, with improvements, could be utilized by the military in a national emergency. In October 1940, Congress appropriated $40 million for the Development of Landing Areas for National Defense (DLAND) program. Funds provided by the DLAND program were utilized at municipal airports in Alabama, including establishment of the Birmingham Army Air Base.

ARMY AIR BASE
BIRMINGHAM, ALABAMA

In 1942, the Army Air Forces began transforming selected municipal airports into Army air bases. The Birmingham Airport was leased to the government in support of national defense. Assigned to the Third Air Force, the airport was utilized as a tactical fighter base. During this period, numerous improvements were completed at the facility. In August 1948, the airport was returned to the City of Birmingham.

114

The Birmingham Modification Center, operated by Bechtel-McCone-Parsons, was established in 1943 at the municipal airport to modify production military aircraft for operation in specific theaters of combat. Modification centers were established to incorporate changes to aircraft designs without disrupting production lines. The first aircraft to be modified at the Birmingham facility was a Consolidated B-24 Liberator heavy-bombardment aircraft christened *Vulcan of Birmingham*.

Consolidated B-24 Liberator aircraft would be flown from production facilities at Willow Run, Michigan, to Birmingham for modification. Upon completion of the work, military pilots or members of the Women Air Force Service Pilots would ferry the aircraft to coastal military air bases for deployment to overseas theaters of combat. The Mobile Air Service Command at Brookley Field was the second major modification center in Alabama.

The Birmingham Modification Center incorporated post-production design changes to the Consolidated B-24 Liberator and Boeing B-29 Superfortress heavy-bombardment aircraft, Douglas A-20 Havoc light bombers, and Lockheed P-38 fighter aircraft. Before the end of the Second World War, workers at the Birmingham facility had modified more than 5,000 aircraft, including 49 percent of the B-29 Superfortress aircraft produced.

By 1944, the Birmingham Modification Center in Birmingham employed 14,000 workers. Because of the labor shortage created by the demands of wartime production, 40 percent of the employees were women. These women worked as mechanics, welders, and aircraft inspectors. In many cases, women workers were preferred for tasks that involved repair of delicate instruments and those performed in confined spaces.

116

Recognizing the success of the use of assault gliders by German military forces in Europe, the Army Air Forces initiated the development of a massive glider-pilot training program. Because the military lacked experience in the operation of gliders, civilian schools were contracted to provide training to military pilots. In July 1942, the Elmira Area Soaring Corporation established a basic flying training school at Bates Field in Mobile. (AFHRA.)

The 18th Army Air Forces Glider Training Detachment was established to provide a four-week course of ground instruction in meteorology, aerodynamics, maintenance, and navigation. Cadets also completed 30 hours of flight instruction with emphasis on proficiency in spot landings, assault operations, and cross-country towing. Upon graduation, the trainee was awarded the rating of glider pilot, promoted to the rank of staff sergeant, and transferred to an advanced school for training in the CG-4A Troop Glider. (AFHRA.)

This Aeronca TG-5 training glider is being towed to altitude over the St. Elmo auxiliary landing field. The field was constructed to relieve aerial congestion during peak periods of training at Bates Field in Mobile. Cadets and instructors utilized the field to practice takeoffs and landings. After the war, the facility was acquired by the State of Alabama through the War Assets Administration for use as a municipal airport.

Military glider pilots also received training in the use of firearms, infantry combat tactics, close order drill, and other military procedures. In combat, the pilot was expected to accompany members of the infantry that were on board the glider until being able to return to base by any means available. Flight Officer Samuel Fine, a graduate of the training program at Mobile, became the first American to land a glider in combat. (AFHRA.)

118

FLIGHT LINE, SOUTHERN AIRWAYS, INC., DECATUR, ALA. 2156AAFBU

The 65th Army Air Forces Flying Training Detachment at Decatur was established as a contract school to provide primary flight training to military cadets. Flight-training operations were conducted by personnel of the Southern Aviation Training School operated by Southern Airways. Founded by Frank Hulse, a native of Birmingham, Southern Airways established military contract training schools throughout the southeastern United States.

This 1940 image shows the facilities of the 65th Army Air Forces Flying Training Detachment. The flying field is visible in the upper portion of the image. The administration buildings, classrooms, and maintenance hangars were located west of the flying field. Primary training consisted of a 10-week course that included flight training, academic instruction, and military procedures. Following graduation, cadets were transferred to a basic flight-training facility. (AFHRA.)

From July 1941 until August 1944, instructors at the Alabama Institute of Aeronautics at Tuscaloosa were assigned to train British and French military aviators. The first class of 70 British cadets arrived in June 1941. The young British aviators were accepted by the local population as extended members of the family. On weekends, cadets attended church, local family functions, and recreational activities planned for their benefit.

To accommodate the increased level of training at Tuscaloosa, the operator of the 51st Flying Training Detachment, Parks Air College of East St. Louis, Illinois, constructed cadet barracks, a dining hall, and academic buildings. The operation utilized three auxiliary landing areas—Rice, Moody, and Knauer Fields. The 51st Flying Training Detachment was the first primary flight school to be established in the Southeast.

Due to the expansion of flight training at Van de Graaff Field, aircraft congestion had become a problem of increasing concern. To alleviate this congestion, Foster Field was constructed adjacent to the Greensboro Highway six miles south of Tuscaloosa. The field was established for use by the Alabama Institute of Aeronautics Civilian Pilot Training Program. The facility was named for University of Alabama president Richard Clarke Foster. (AAHOF.)

In 1939, Nancy Batson (left) became one of five females students enrolled in the Civilian Pilot Training Program at the University of Alabama. One of 28 original members of the Women's Auxiliary Ferrying Squadron, Batson became the first Alabama female to pilot a military aircraft. From 1942 to 1944, Batson flew numerous single- and multiengine military aircraft. In 1989, Batson became the first female to be inducted into the Alabama Aviation Hall of Fame. (AAHOF.)

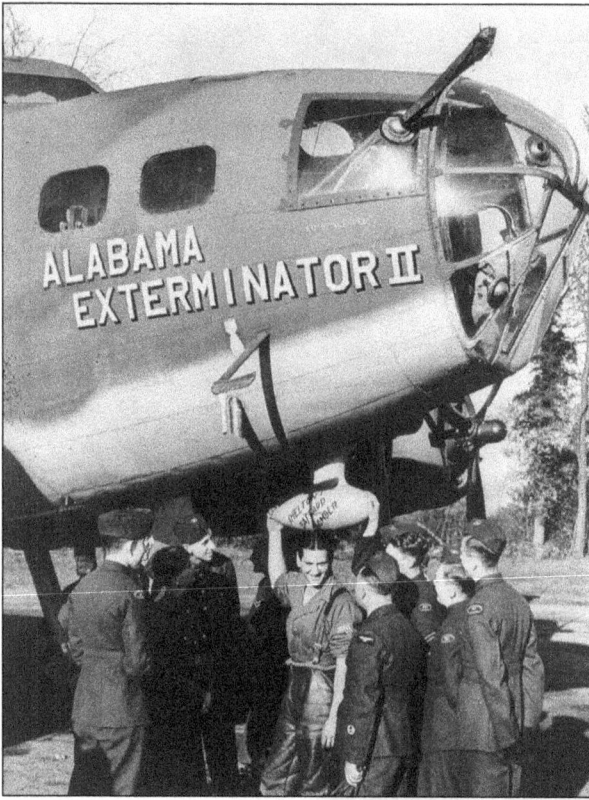

The *Alabama Exterminator II* was one of a number of military aircraft utilized during the Second World War named in part for the state of Alabama. The aircraft was a Boeing B-17F heavy-bombardment aircraft assigned to the 384th Bomb Group in England. The original B-17, *Alabama Exterminator*, was destroyed in a crash landing in June 1942 on a rocky beach in Greenland.

A native of Bessemer, David McCampbell (1910–1996) was a US Navy captain, fighter pilot, and Medal of Honor recipient. McCampbell is the all-time leading flying ace of the United States Navy, with 34 aerial victories, and the third-highest American scoring ace of the Second World War. He also set a single-mission aerial combat record in October 1944 by destroying nine enemy aircraft. (AAHOF.)

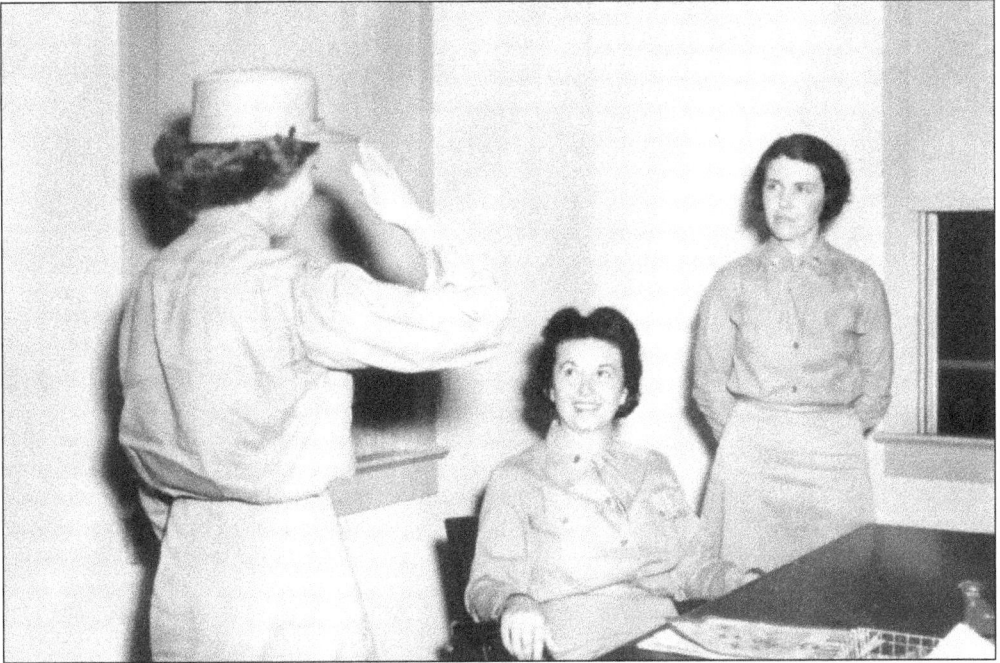

Women served a vital role in support of the United States military during the Second World War. The Women's Army Auxiliary Corps, established in 1942, consisted of women trained to serve in noncombat positions. In July 1943, the program was redesignated as the Women's Army Corps. The first contingent of 156 women arrived at Maxwell Field in April 1943 to serve as clerks, stenographers, and technicians. (AFHRA.)

In December 1943, the first members of the Women Airforce Service Pilots entered service at Maxwell Field. Members served as copilots in Consolidated B-24 heavy-bombardment training aircraft at Maxwell Field. Detachments also served at Gunter Field in Montgomery, Napier Field at Dothan, and Craig Field in Selma conducting flight tests of numerous types of training aircraft and daily inspections of auxiliary landing fields. (AFHRA.)

Following the victory of Allied forces in Europe and Asia, military training airfields in Alabama were rapidly decommissioned. Often, the wooden barracks and other structures were removed and reassembled by local residents for use as homes and utility buildings. The runways of these once proud facilities were subsequently used by automobile clubs for drag races, road races, and other automotive sporting events.

Many former military airfields in Alabama were repurposed for use as municipal airports or industrial development. By the year 2000, the former Tuskegee Army Air Field had virtually disappeared, planted in longleaf pine trees for future harvesting. Moton Field, former site of the 66th Army Air Forces Flying Training Detachment, continues to serve as a municipal airport and has been designated Tuskegee Airmen National Historic Site.

During the Second World War, the Anniston Army Air Field at Eastaboga was utilized as an auxiliary landing facility for Boeing B-24 and B-29 heavy-bombardment aircraft operating from Maxwell Field and training fields at Courtland and Smyrna, Tennessee. The field was also utilized as a storage and retrofit facility for the modification center at the Birmingham Municipal Airport.

After the war, the former Anniston Army Air Base served as the municipal airport at Talladega. In 1969, the International Speedway Corporation transformed the field into a motorsports complex known as the Alabama International Motor Speedway. The concrete, steel-reinforced runways became part of the racetrack complex. The hangar, operations building, and other structures located on the base are utilized for manufacturing and other commercial purposes.

The once modern structures of former municipal airports and military airfields are reminders of the remarkable aviation heritage of the state of Alabama. The abandoned administration building of the original Montgomery Municipal Airport is located in the Gunter Industrial Park. The former municipal hangar, erected in 1929, is visible to the rear of the administration building.

The conclusion of the Second World War did not bring an end to flights of heavy-bombardment aircraft over Alabama. In 1948, scenes of the movie *Twelve O'Clock High*, starring Gregory Peck, were filmed at an abandoned military airfield at Ozark. Written by Lt. Col. Sy Bartlett and Col. Beirne Lay Jr., the movie was produced with the assistance of personnel from Maxwell and Eglin Fields. (AFHRA.)

Many former military airfields in Alabama became local municipal airports through the Surplus Military Airfield Program. Pryor Field in Decatur was named for Schuyler Harris Pryor, who lost his life while serving as a flight instructor at the facility. Deactivated in December 1944, the facility was transferred to the City of Decatur by the War Assets Administration. In 2010, the airport was listed in the Alabama Register of Landmarks and Heritage.

The parcel of land used by Orville and Wilbur Wright to establish the nation's first civilian flying school has maintained a continuing presence during the first century of powered flight in Alabama. First constructed to support flight training operations in the Southeast during the First World War, Maxwell Air Force Base is currently the center for aerospace education and the development of airpower doctrine for the United States Air Force. (AFHRA.)

Visit us at
arcadiapublishing.com

www.ingramcontent.com/pod-product-compliance
Lightning Source LLC
Chambersburg PA
CBHW080547110426
42813CB00006B/1234